METABIOTICS

Boris A. Shenderov • Alexander V. Sinitsa
Mikhail M. Zakharchenko • Christine Lang

METABIOTICS

PRESENT STATE, CHALLENGES AND PERSPECTIVES

Springer

Boris A. Shenderov
Research Laboratory for Design
& Implementation of Personalized
Nutrition-Related Product & Diets
K.G. Razumovsky University of Technology
& Management
Moscow, Russia

Alexander V. Sinitsa
Kraft Ltd.
St. Petersburg, Russia

Christine Lang
MBCC Group
Berlin, Germany

Mikhail M. Zakharchenko
Kraft Ltd.
St. Petersburg, Russia

ISBN 978-3-030-34166-4 ISBN 978-3-030-34167-1 (eBook)
https://doi.org/10.1007/978-3-030-34167-1

This Springer imprint is published by the registered company Springer Nature Switzerland AG
The registered company address is: Gewerbestrasse 11, 6330 Cham, Switzerland

Preface

This book presents in a concise form the latest data on symbiotic microbiota which is directly or indirectly involved in the evolvement and performance of human physiological functions, biochemical, behavioral, regulatory, and signaling reactions, in maintaining energy, immune, and neurohormone homeostasis. Multifaceted contribution of this microbiota in health support is ensured by endogenous production of multiple low-molecular-weight microbial compounds, many of which bear similarity to biologically active molecules of cells of human tissues and organs in their chemical structure, pharmacological and signaling activity, or which can be found in food products. Their lack or disproportionately large amount results in imbalance of mitochondrial, microbial, and cell metabolites and may become a predictor and probably the main factor of metabolic diseases. This leads to the conclusion that microbial low-molecular-weight bioactive compounds are the most important human health regulators in all periods of life and the risk of many chronic metabolic diseases depends on adequate diet and balanced condition of digestive tract microbiota. The most common approach to preserving human symbiotic microbiocenoses consists in using probiotics and prebiotics. Unfortunately, beneficial effects of probiotics on the base of live microorganisms are often short-term, uncertain, or entirely absent; the application of traditional probiotics may also cause various side effects. The book offers a fundamentally new approach to the prevention of chronic deficiency of biologically and pharmacologically active low-molecular-weight microbial compounds by way of introduction into preventive and clinical medicine of functional nutrition products (metabiotics) engineered on the basis of cell structural components, metabolites, and signaling molecules of probiotic microbial strains with a determined (known) chemical structure. Metabiotics are capable of optimizing host-specific physiological functions, metabolic, epigenetic, informational, regulator, transport, and/or behavior reactions connected with the activity of symbiotic microbiota. They can act as therapeutics or as nutritional supplements to traditional functional foods. In the future, as is the case with antibiotics, synthetic (and/or semisynthetic) metabiotics may be expected to appear, which will be artificially created as analogues or improved versions of natural biologically active compounds formed by symbiotic microorganisms. It should be borne in mind that many issues related to understanding the interaction between metabiotics and host organism remain open. This publication is intended for broad audience including doctors of all categories, dieticians, biochemists, food technologists, teachers, and students of higher educational institutions or departments specializing in medicine, biology, and foods, as well as for all those interested in drug-free healthcare methods.

UDC 615
LBC 52.64
Shenderov BA, Sinitsa AV, Zakharchenko MM, Lang C
Metabiotics: Present State, Challenges, and Perspectives

Moscow, Russia — Boris A. Shenderov
St. Petersburg, Russia — Alexander V. Sinitsa
St. Petersburg, Russia — Mikhail M. Zakharchenko
Berlin, Germany — Christine Lang

Contents

About the Authors

Boris A. Shenderov is a physician, doctor of Medical Science (1976), professor of Microbiology (1979), active member of New York Academy of Sciences (1996). Also, he is professor in Scientific Research Laboratory "Design and Implementation of Personal Nutrition Products and Rations," K.G. Razumovsky Moscow State University of Technologies and Management, and in Interdisciplinary Neurology Department, Institute of Interdisciplinary Medicine, Moscow, Russia. His main research interests include medical microbial ecology, genetics, epigenetics, metabolomics, probiotics, functional foods, and biotechnology. He is an author of more than 500 scientific papers, including 12 books, and 30 invention patents. He was a councilor of Executive Board of International Society of Microbial Ecology and Disease (SOMED) (2008–2013) and a member of Executive Board and President of International Association for Gnotobiology (IAG) (2008–2014); Russian Society for Epidemiology, Microbiology, and Parasitology after I.I. Mechnikov; and Editorial Boards of five Russian science journals and international journal *Microbial Ecology in Health and Disease*.

e-mail: shenderof@yandex.ru (professional)

Alexander V. Sinitsa is president of "Kraft" Group, Saint Petersburg, Russia, and candidate of Technical Sciences (1987). His main scientific interests include probiotics, prebiotics, metabiotics (postbiotics), functional foods, and sports nutrition. He is an author and coauthor of more than 100 scientific papers. He heads a company, one of the leaders in the development of metabiotic products in Russia.

e-mail: prezident@kkraft.ru (professional)

Mikhail M. Zakharchenko is a physician, candidate of Medical Science (2003), and head of R&D Department of "Kraft" LLC, Saint Petersburg, Russia. His main scientific interests include microbiome, probiotics, prebiotics, metabiotics (postbiotics), and functional foods. He is an author and coauthor of more than 100 scientific papers and a member of Executive Board of Scientific Conference "Prenosology" (2005–2019) and Editorial Board of Russian Science Journal *Prenosology and*

Healthy Lifestyle (2007–2019). He leads the development of new products at "Kraft."

e-mail: m.zaharchenko@kkraft.ru (professional)

Christine Lang is an entrepreneur in biotechnology. She acts as a consultant for the German government for bioeconomy topics and co-chairs the German Bioeconomy Council. She is an associate professor of Microbiology and Molecular Genetics at Technical University of Berlin and vice president of the German Association of General and Applied Microbiology (VAAM). In 2001, she founded the company Organobalance GmbH and established it as an R&D facility to develop novel microbial active ingredients and strains for food, feed, personal care, and pharma applications. Since September 2016, Organobalance is a part of Novozymes A/S. Till June 2018, she was the company's CEO and now serves as a consultant for Novozymes A/S. Christine Lang is co-founder and partner of BELANO medical AG focused on production and sales of microbiotic products for human health.

e-mail: christine.lang@mybioconsulting.de (professional)

Introduction

All diseases start in the gastrointestinal tract.

(Hippocrates 460–370 BC)

Multiple and numerous associations of microorganisms inhabiting human gastrointestinal tract determine, to a great extent, our physical and spiritual health.

(Mechnikov 1908)

Microbes rule the world. We must listen closely to their language and when we learn more, we will be capable of a better and more harmonic interaction with them. The presence of active microbial compounds in the gut has physiological and pathophysiological consequences for the host.

(Midtvedt 2008)

Microorganisms are starters of the nascence and subsequent evolution of all biological life varieties on our planet including humans. A contemporary view of a human organism presents it as a superorganism, a dynamic symbiotic community of various prokaryotic and eukaryotic cells (Vakhitov and Sitkin 2014, Shenderov 2008, Shenderov 2014b, Ugolev 1991, Caselli et al. 2011, Dietert and Dietert 2015, Lederberg 2000). The microbial component of this community includes bacteria (which dominate), *Eukarya* and *Archaea* (up to 10^{13-14} cells totally). Microbiome/microbiota (a term introduced into scientific literature by *Joshua Lederberg*) represents the totality of commensal, symbiotic and pathogenic microorganisms inhabiting human body (Lederberg and McCray 2001). According to the latest data, the human organism contains around 3×10^{13} eukaryotic cells and 3.9×10^{13} microorganisms colonizing human body; it means that quantitatively, the content of human cells and microorganisms is quite similar (Sender et al. 2016). As regards bacterial content, symbiotic microbiota in the digestive tract reaches its maximum in the large intestine (up to 10^{12} CFU/g); this is the most densely inhabited bio-ecosystem on our planet. Human microbiota has a distinctly individual character on the genus

B. A. Shenderov et al., *METABIOTICS*,
https://doi.org/10.1007/978-3-030-34167-1_1

and species levels and — especially — on the strain levels (Gilbert et al. 2016, Meisel and Grice 2017, Mimee et al. 2016, Sonnenburg and Backhed 2016).

Multiple and numerous intestinal microorganisms are the main adaptive component of the human superorganism determining close mutually beneficial relations between the metagenome, meta-epigenome and intercellular information exchange of all its participants. It ensures adaptation of the human organism to ever-changing environmental factors, maintaining its energetic, metabolic and immune homeostasis. Moreover, the digestive tract, skin and vaginal microbiota has diverse effects on the development and functioning of the nervous and the hormone systems, on social and psychical behavior. The structure of symbiotic microbial population in women prior to, during and after pregnancy influences the development of the fetus, the genesis and succession of fetal and infant microbiota, and even the course of pregnancy of their children and their children's offsprings. The connection has been traced between the microbiota of pregnant and breastfeeding women and oligosaccharide content in breast milk; the quantitative and qualitative composition of milk oligosaccharides determines, to a large extent, the development and content of fetal microbiota and that of newborn babies and infants, and in the long run — the health of teenagers and adults (Charbonneau et al. 2016). It means that indigenous microbiota of all mammals including man is a complex, indispensable, peculiar to them extracorporeal organ, which plays a fundamental role in maintaining their health and reducing the risk of various metabolic diseases. Practically all functions, metabolic and signaling reactions in the cells of human organs and tissues are in accord with the content and functions of its symbiotic microbiota. It is beneficial for both the whole organism and its components. The effects of symbiotic (probiotic) microorganisms are realized through the production of multiple low molecular weight compounds of different nature and chemical structure (Shenderov 2013c, Shenderov 2014b, Cani 2018, Shenderov 2011a, Shenderov 2013a, Sonnenburg and Backhed 2016).

The "host – its normal microbiota" ecosystem bears innate self-regulatory elements and is capable to withstand – at least, within bounds – environmental changes and abrupt fluctuations in microbial populations density. Unfortunately, many biological and abiotic factors and agents are capable of damaging natural symbiotic microbiocenoses and their interaction with eukaryotic cells of human tissues and organs, which is a predisposing risk factor for not only traditional infectious diseases but also for many somatic metabolic diseases associated with microecological imbalance. Many various therapeutics, dietary supplements, functional foods have been developed and are used in order to preserve and restore human microbial ecology. The most popular among them are those based on the use of specially matched probiotic strains of lactobacilli, bifidobacteria and some other microorganisms, as well as soluble dietary fibers stimulating their growth. The efficiency of traditional microecological therapeutics (probiotics, prebiotics, synbiotics) that dominate nowadays and are used for maintaining and restoring human symbiotic microbiota is determined by the origin of probiotic cultures (hetero-, homo- or autoprobiotics), the kind of species and strain they belong to, the state of probiotic bacteria (live, killed, cell fragments, their metabolites or signaling molecules), the probiotic manufacturing technologies, the method and duration of use, the quantity of viable probiotic microorganisms, the user's health status, physical and chemical conditions in different biotopes of his gastrointestinal tract, the state of gut microbiota, the combination of all the above-mentioned and other factors. Given that, probiotics display their direct or indirect, specific or nonspecific effects — both on a local level and on the systemic level, as a result of a long-term or transitory colonization of the host organism.

Ample data have been published concerning probiotics based on live organisms highlighting their health-promoting effects in case of acute and chronic, localized and systemic diseases (various types of diarrhea, allergy, inflammatory pathologic conditions in the gut, hypercholesterolemia, malignancies etc.) (Shenderov 2014b, Bomba et al. 2012, Pflughoeft and Versalovic 2012, Shenderov 2011b, Sonnenburg and Backhed 2016). In this book, we have analyzed the materials published over the last 10 years both in Russia and globally regarding positive effects, drawbacks and adverse consequences of ample use of probiotics on the base of live microorganisms in medicine and veterinary practice. The concept of probiotics is being revised now. It is giving way to the concept of metabiotics whose main constituent is represented by cell components, metabolites and signaling molecules of probiotic cultures. The presented materials summarize the data available in scientific literature on the such microbial molecules, new approaches are considered to constructing safe and effective microecological therapeutics on their basis, and new data is presented on some of the best understood molecular mechanisms of metabiotics' positive effects on human organism.

The Composition and Functions of Human Gut Symbiotic Microbiota

From the contemporary perspective, the human organism should be viewed as the most complex "superorganism", a symbiotic community of eukaryotic, prokaryotic cells including archaebacteria, and viruses (Ugolev 1991, Lederberg 2000). The microbial component of this community is represented by the aggregate of sets of microbiocenoses characterized by a definite composition and occupying the respective biotope in the human organism which is open to the environment (skin, nasopharynx, mouth cavity, respiratory and gastrointestinal tracts, genitourinary system mucosa). In any microbiocenosis one can distinguish widespread species, the so-called characteristic or dominating (*core*) species (autochthonous, indigenous symbiotic microbiota) and additional or accidental species (transitory allochthonous microbiota). The number of characteristic species is relatively not too big, but to make up for it, they are always well-represented. The "metagenome" of this "superorganism" consists of the *Homo sapiens* genes proper and the genes of microorganisms colonizing human body. The genome of every human being is quite stable (except for the changes in genes related to the immune system, the metabolism of various dietary substrates or xenobiotic destruction, neoplasms); the microbiome, on the other hand, undergoes rather profound changes in the course of a lifetime (Gilbert et al. 2016).

The last twenty years have seen a sharp increase in the information flow on the important role played by symbiotic microbiota in maintaining human physical and mental health. The example of gut microbiocenosis shows the existence of a complex ramified and multi-tier system for cooperation between populations of microorganisms inhabiting the gut, as well as between these microorganisms and eukaryotic cells of the gut and human organism as a whole. Deficiency or excess of a certain substrate, metabolite or signaling molecules serves as a signal for increased growth, death or disfunction of the corresponding chain of the complex system called "human superorganism." In course of evolution, symbiotic microorganisms and cells of plants and animals including humans had been transforming into an increasingly interconnected whole. For greater efficiency, they had been specializing as regards the habitat and functional activity. This kind of integration has made

B. A. Shenderov et al., *METABIOTICS*,
https://doi.org/10.1007/978-3-030-34167-1_2

it possible to consider symbiotic and commensal microbiota as a peculiar extracorporeal human organ, which works concertedly for the benefit of the whole system of the host organism where it is localized.

Russian academician A. M. Ugolev (1991) was the first to formulate the notion of endo-ecology of humans and other higher organisms, having pointed out that from the trophic and metabolic viewpoint, mammals represent a most complex system consisting of the dominating multicellular organism and specific bacterial cultures maintaining complex symbiotic relations. All the necessary plastic, energetic, metabolic and signaling regulatory compounds are released from food products due to cavitary, membrane, intracellular digestion, as well as in course of their synthesis by gut microorganisms. Regulatory substances (the so-called exohormones) may be present in food or be produced from it under the effect of digestive enzymes of the microorganism and a multitude of enzymes of various bacteria colonizing its skin and mucosae. Such exohormones are necessary for functioning of the organism as a whole, not only the digestive system.

In subsequent years, a number of amendments were introduced into A. M. Ugolev's view of human superorganism, the most important of which is about humans representing a community of prokaryotic cells rather than eukaryotic cells (Sitkin et al. 2015, Shenderov 2014b, Dietert and Dietert 2015, Lederberg 2000). This conclusion is based on the fact that various kinds of cells of specifically human tissues and organs do not exceed 500 in number; nevertheless, the total number of these cells ranges between 5 and 10 trillion. On the contrary, the number of various species of bacteria inhabiting the human organism may reach 10 thousand, and the number of strains may be as much as 50 thousand; the total bacterial content is within hundreds of trillions, and with viruses included it exceeds quadrillion. The number of genes in human chromosomes is in the range of about 25,000, while the human microbiome includes up to 10 million genes. The replacement of all eukaryotic cells in the human being requires not less than 20–25 years. During this time, all symbiotic bacteria can be replaced at least five or six times, which is indicative of high adaptive capacity of a human being as a superorganism. Up to 80% of all man's energy is produced in mitochondria (the most ancient microbial endosymbionts of eukaryotic cells), 20% of energy is given by gut microorganisms. Due to the metabolic activity of gut bacteria, over 90% of energy is generated for the epithelial cells of the digestive tract (Shenderov 2016a). Over 70% of prokaryotic organisms are obligatory anaerobes; and now only the representatives of a little more than 1500 bacterial species can be cultivated. The representatives of 50– 60% of bacterial species in the gut produce spores, which is believed to help obligatory anaerobe bacteria persist and get transmitted to other hosts (Braune and Blaut 2016, Ilinskaya et al. 2017).

Long-term studies of gut microbiota composition in one and the same persons have shown that 60% of strains of dominant bacterial species are preserved in this biotope for not less than 5 years (Faith et al. 2013, Nicholson et al. 2012). Such long-term individual stability of gut or skin microbiota which allows for identifying up to 80% of people by their microbiota content has given grounds to speak about the existence of unique microbial fingerprints (Franzosa et al. 2015). The large intestine is predominately inhabited by such phylum representatives as *Bacteroides*

Table 1 The composition of digestive tract microbiome in healthy people (Shenderov 2014b, Shenderov et al. 2018a, Bik et al. 2018, Blum 2017, Chernevskaya and Beloborodova 2018, Falony et al. 2016, Meisel and Grice 2017, Yadav et al. 2018, Zhernakova et al. 2016)

Phylum	Genera
Firmicutes	*Blautia, Butyrivibrio, Clostridium, Coprococcus, Dorea, Eubacterium, Faecalibacterium, Roseburia, Ruminococcus, Subdoligranulum; Bacillus, Lactobacillus, Streptococcus*
Bacteroides	*Bacteroides, Prevotella, Aliistipes*
Actinobacteria	*Bifidobacterium, Collinsella*
Fusobacteria	*Fusobacterium*
Proteobacteria	*Desulfovibrio, Bilophila, Esherichia*
Verrucomicrobia	*Akkermansia*
Melainabacteria	*Cyanobacteria*
Euryarchaeota	*Methanobacteriales, Metanomassiliicoccales*
Fungi	*Saccharomyces, Candida, Cladosporum*
Viruses	*Eukaryotic viruses, Bacteriophages*

and *Firmicutes* followed by *Actinobacteria, Proteobacteria, Fusobacteria, Spirochaetes, Verrucomicrobia, Lentisphaerae* and *Archea* (Table 1).

The ratio between these phyla changes throughout life. For example, in infants, adults and people older than 75 the ratio between the representatives of *Firmicutes* and *Bacteroides* in the large intestine content is 0.4, 10.9 and 0.6 respectively. Only 18– 20 bacterial species are ever-present in the majority of adult people. The representatives of 14 gut microbial genera (*core microbiota*) are found in the majority of 4000 screened inhabitants of Eastern and Western Europe; given that, the representatives of other 664 genera are barely identified and are found with different incidence in adults (Falony et al. 2016). Human microbiota has a distinctly individual character and varies both on the level of the genus and species and on the strain level. Intraspecies differences among strains can reach 25% of their genome and more. Bacterial content (CFU/g) and the number of species among particular individuals can vary 12–2200-fold. Fecal bacteria differ in viability: 30% of them are dead cells; 50% are live cells, and 20% are damaged cells with a low capacity to form colonies on growth media (Shenderov 2014b, Carding et al. 2015, Charbonneau et al. 2016, Gilbert et al. 2016, Sonnenburg and Backhed 2016, Suvorov 2013, The Human Microbiome Project Consortium 2012). Also, over 1200 species of viruses, mostly bacteriophages, are present in the gut. Over recent years, increasingly more attention is paid to fungi, protozoans and helminths in the framework of research into human microbiome (Proal et al. 2017). Feeding habits, natural surroundings, immune and genetic peculiarities of the host are the decisive factors for the composition and functions of indigenous microbiota of human digestive tract and its metabolic phenotype (Donaldson et al. 2016, Nicholson et al. 2012, Voreades et al. 2014, Yadav et al. 2018). According to the proportion of microbial phyla representatives (*Bacteroides, Prevotella* and *Ruminococcus*) in the large intestine microbiota, the majority of people can be divided into three “enterotypes.” Russian urban and rural people differ in the number and content of bacterial "enterotypes" in their digestive tract (Vakhitov and Sitkin 2014, Sitkin et al. 2015).

Owing to a well-coordinated work of symbiosis-regulating systems that have taken shape in course of the evolution, the genes of eukaryotic cells and microbiomes of many thousands of symbiotic microorganisms are in a close and consistent interaction. In the long run, this is what determines the composition and quantitative content of the corresponding prokaryotic and eukaryotic cells in particular biotopes of mammals including humans, which prevents their inevitable competition for similar fixation places and dietary substrates, regulates the exchange of metabolites, signal molecules, genetic information transmission etc. Symbiotic microorganisms are a source of a multitude of endogenous mono- and multiple-factor signaling molecules that ensure health or risk of human diseases from the moment of birth till extreme old age (Sitkin et al. 2015, Chervinets et al. 2018, Shenderov 2014b, Carding et al. 2015, Gilbert et al. 2016, Lagier et al. 2012, Lyu and Hsu 2018, Sonnenburg and Backhed 2016). For the most part, symbiotic microorganisms exist in human and animal organisms as microcolonies fixed to particular receptors and enclosed in a biofilm, which provides a glove-like cover to the skin and mucosae lining up the cavities open to environmental influence in healthy humans and animals. Apart from microorganisms, this biofilm contains exopolysaccharides of microbial origin varying in composition, as well as mucin produced by mucosal goblet cells. From the functional perspective, biofilm is often compared to placenta. While the latter regulates the relations between the fetus and the mother's organism, the biofilm plays the same role in the relations between the host organism and the environment. Dietary peculiarities by 57% determine structural changes in gut microbiota composition; genetic predisposition is only by 12% responsible for such peculiarities of microbiota. There are dozens some population groups among people that differ greatly as regards their diets (innate inhabitants of tropical and subtropical zones, deserts, highlands, northern territories, strict vegetarians, people of the so-called "western pattern diet"). Similarly, commensal and symbiotic microbiota of the digestive tract in the representatives of these ethnic groups is characterized by considerable or even critical differences (Shenderov 2014b, Sonnenburg and Backhed 2016).

In all multi-cellular organisms including humans, symbiotic microorganisms are actively involved in physiological functions, biochemical, behavioral and signaling reactions, in maintaining health and in development of various, primarily metabolic, diseases. Immune, metabolic and signaling functions of symbiotic microbiota similar to those of adults, take shape by the age of 2 or 3 years, developing to the full extent by the age between 12 and 14. Therewith, metabolic changes coincide with the evolution and maturation of gut symbiotic microbiota primarily on the strain level, which is confirmed by studying and comparing the profiles of bacterial products of protein and energy metabolism in different human and animal body fluids (Nicholson et al. 2012). Under standard conditions, normal microbiota effects directly or indirectly practically all vitally important human processes and functions (morphokinetic action; regulation of gas composition in the cavities and water-salt exchange; participation in the metabolism of proteins, lipids and carbohydrates;

supplying the body with energy, in the recirculation of bile acids and other macromolecules; immunogenic and detoxication functions; mutagenic/antimutagenic and oxidative/antioxidative activities; regulation of behavioral reactions; the storage of genetic material; production of low molecular weight compounds of different chemical nature, with a broad spectrum of biological, pharmacological and signaling activity; regulation of metagenome stability; replication and phenotypic gene expression, modification of genomic and post-translational reactions of prokaryotic and eukaryotic cells, programmed death of eukaryotic cells (apoptosis); involvement in information exchange between prokaryotic and/or eukaryotic cells of the host, in disease etiopathogenesis and functioning as starter cultures when creating various biotechnological products) (Gao et al. 2013; Shenderov 2014b, Cani and Delzenne 2007, Dorrestein et al. 2014, Freitas et al. 2003, Maguire and Maguire 2019, Lyu and Hsu 2018, Mimee et al. 2016, Nicholson et al. 2012, Yadav et al. 2018).

Contemporary Views on Biotechnological Potential of Symbiotic Microorganisms

Microorganisms are essentially the main "workhorses" of contemporary biotechnology. It should be noted that a large proportion of microorganism strains existing in nature, including many representatives of anaerobe symbiotic bacteria in the human gut, are difficult and often impossible to cultivate in ordinary microbiological laboratories (Browne et al. 2016). As many as 1000–10,000 prokaryotic microorganism species are present in 1 g of soil. Out of 23,000 microbial metabolites known to date (42% of them are produced by fungi and 32% by actinomycetes), many are used already in the form of antibiotics, anti-cancer agents, immunosuppressors, hypocholesterolemic, antivirus, deworming drugs, nutraceutics, dietary and technical exopolysaccharides, surfactants, herbicides, various enzymes, amino acids, vitamins, vaccines etc. Suffice it to say that over 15,000 various natural and synthetic and semi-synthetic antibiotics created on their base are known by now, and 150 of them are commercially available. To date, over 1000 antimicrobial peptides have been isolated, with bacteriocins being a common example. The representatives of the *Myxobacteria* genus alone form over 300 different antimicrobial compounds. Presently, the world's market of antibiotics exceeds USD 35 billion (Guilherme et al. 2006). The annual production of the glutamic acid by using microbial starter cultures exceeds 2.2 million tons, lysine — 1.3 million tons, threonine — 77,000 tons, phenylalanine — 16,000 tons, B12 vitamin — 12 tons. The production of the polylactic acid of microbial origin does not differ in its volumes from polypropylene and reaches 140,000 tons a year; the main characteristic feature of the polylactic acid is its capability of quick biodegradation.

In the last 50–75 years, secondary metabolites of microbial origin allowed humankind to achieve a practically double increase in average life expectancy. The overwhelming majority of microbial biotechnological products manufactured worldwide were mostly based on the cultures of microorganisms isolated from soils and waterbodies. In the last decade, representatives of symbiotic microbiota of plants, lower and higher animals including humans came into use for biotechnological purposes (Chervinets et al. 2018). It became the revolution in contemporary biotechnology. Thus, it turned out that symbiotic bacteria of sponges alone make up

B. A. Shenderov et al., *METABIOTICS*,
https://doi.org/10.1007/978-3-030-34167-1_3

to 40% of their biomass. The symbiotic bacteria of these protozoans form up to 400 different secondary metabolites with marked pharmacological and other effects; a number of antineoplastic drugs have been formulated on their basis, two of which have been introduced into clinical practice. It should be remembered that the skin and mucosa of an adult carries up to 1.5–3 kg of commensal and symbiotic microorganisms belonging to over 1500 genera and including up to 40–70 thousand different strains.

In 2011, the world's market of traditional probiotics and probiotic fermented milk products on the base of symbiotic lactic acid bacteria of animal origin and prebiotics was valued at USD 27.9 billion, in 2015 it was valued at USD 30 billion; it is expected to reach USD 44.9 billion in 2018, with a yearly growth between 2 and 5% (Buriti et al. 2016). According to *Elisa Fernandez*, director of the *International Probiotics Association (IPA)*, the world's probiotics market is expected to reach EUR 53 billion by 2023 (www.internationalprobiotics.org). It should be borne in mind that a number of commensal bacteria (representatives of *Ruminococcus, Eubacterium, Roseburia, Faecalibacterium, Akkermansia spp.* etc.) that do not belong to traditional probiotic bacteria, can also produce various bioactive metabolites capable of beneficial effects on the human organism (Engevik and Versalovic 2017). Probiotic strains commercially available on present-day markets are just the first generation of biotechnological products. In the perspective, symbiotic microorganisms of mammals will become the base for starter cultures in the production of the newer, safer and more effective therapeutics that restore or improve the host organism microecology (Chervinets et al. 2018, Shenderov et al. 2018a).

The Digestive Function of Human Gut Microbiota

According to the contemporary data, the daily needs of the human organism required for construction and functioning of millions of simple and complex molecular entities amount to over 20 thousand of different macro- and micronutrients (Shenderov 2018). Around 70% of the total microbiota inhabiting the human organism is localized on its digestive tract mucosa. A. M. Ugolev deserves credit for providing convincing evidence of symbiotic microorganisms actively participating in the metabolization of complex compounds (polysaccharides, phenolic and other components) coming with food. In recent years, the importance of the digestive function of symbiotic gut bacteria has been gaining further confirmation (Bengmark 1998; Bik et al. 2018, Sonnenburg and Backhed 2016, Yadav et al. 2018). By digesting endogenous sources, gut microbiota returns all the necessary components. It has been proven that over 400 or 450 grams of endogenous substrates and those coming with food undergo daily microbial metabolization in the digestive tract of adults, which makes up as much as one third of the total mass of the consumable food products (Table 1).

Form the chemical perspective, human "superorganism" is supposed to consist of not less than 2.5 million molecules including around one million proteins (Wang et al. 2006), 300 thousand lipids (Sampath and Ntambi 2005) and hundreds of thousands of other simple and complex compounds (Engevik and Versalovic 2017). Unlike symbiotic microbiota and its microbiome in people living in industrial countries, the metabolome of ethnic groups has not been extensively characterized until now, which is explained by its extreme diversity (Dorrestein et al. 2014). Formerly, food products were believed to be the only basic substrates, co-substrates and co-factors for the synthesis of these compounds. Notably, complex dietary compounds undergo destruction under the influence of the digestive tract enzymes; the resulting compounds of simple chemical structure are absorbed in the gastrointestinal tract and are used by cells for the synthesis of energy and the required structural and signaling molecules. They are either ingested with food products or produced endogenously by various cells of the microorganism, or result from the metabolic activity of symbiotic microbiota (gut microbiota in the first place). Modern diet

B. A. Shenderov et al., *METABIOTICS*,
https://doi.org/10.1007/978-3-030-34167-1_4

Table 1 Dietary and endogenous substrates metabolized daily by gut microbiota in adults (calculated data) (Shenderov 2017)

Substrates	Daily amount
Mucosal fluid (in the first place, mucopolysaccharides of the nasopharynx and gut mucous layer)	Up to 1000 ml (20–25 g of solids)
Saliva	Up to 1.5 l (20–35 g of solids)
Gastric juice	Up to 2.5 l (20–25 g of solids)
Bile	0.5–1.0 l (20–25 g of solids)
Pancreatic juice	Up to 1.0 l (10–15 g of solids)
Small intestine juice	Up to 2.5 l (30–40 g of solids)
Large intestine juice	50–70 ml (1.5–2 g of solids)
Intake of microorganisms with food, water, inhaled air	Up to 10^{11-12} cells (2.5–3 g)
Dead cells of the digestive tract symbiotic microbiota	50–150 g
Desquamated cells of dead epithelium of the digestive tract	10–50 × 10^7 cells (up to 10 g)
The quantity of endogenous substrates metabolized by gut microorganisms	Up to 200–250 g on the average
The quantity of non-digestible dietary substrates metabolized by gut microorganisms	Up to 100–120 g on the average
Total quantity of dietary and endogenous substrates daily metabolized by microbes	Up to 400–450 g

cannot provide the intake of even a half of this amount. In case of dietary intake deficiency, it is compensated by microbial formation of bioactive compounds from various sources of dietary, endogenous and microbial origin.

Gut microbiota is most actively involved in dietary metabolism by processing dietary and endogenous substrates and forming low molecular weight nutrients, regulatory and signaling molecules required for the life sustaining activity of the host and its microbiota. Gut bacteria are capable of splitting various ingested vegetable components (polyphenols, polysaccharides, oligosaccharides etc.) into biologically active molecules which take an active part in different human functions and reactions (Volkov et al. 2007, Moreira et al. 2017; Panyushin 2012). The amount of macro- and microelements (mg/kg) contained in raw biomass of fecal microorganisms in adults exceeds daily requirement by 5–50 times (Shenderov 2008). It is this daily recirculation that helps to partially or fully cover chronic deficiency in many critical nutrients and other biologically active substances of dietary and microbial nature.

Metabolic Relationship Between the Host and Its Gut Microbiota

Symbiotic microorganisms ever-present in the organisms of adult people form over 25 thousand different biologically and pharmacologically active compounds. With regard to potential biological effects, the most extensively studied are short-chain fatty and bile acids, choline metabolites, the derivatives of phenol, benzene and phenyl, indole derivatives, vitamins, polyamines, lipids, enzymes and other proteins, amino acids (Beloborodova et al. 2011, Shenderov et al. 2010, Carding et al. 2015, Chernevskaya and Beloborodova 2018, Clarke et al. 2014, Engevik and Versalovic 2017, Ilinskaya et al. 2017, Maguire and Maguire 2019, Nicholson et al. 2012, Neis et al. 2015, Shaikh and Sreeja 2017), catecholamines and other neuromodulators (Oleskin et al. 2016, Clarke et al. 2014, El Aidy et al. 2016, Ilinskaya et al. 2017, Maguire and Maguire 2019, Oleskin et al. 2017), gas molecules (Shenderov 2015, Carding et al. 2015) and many others (Chervinets et al. 2018, Carding et al. 2015, Engevik and Versalovic 2017, Nicholson et al. 2012). In healthy people, gut symbiotic microorganisms are the source of these microbial bioactive molecules. In the process of evolution, humans have been selecting and retaining the kinds of microorganisms that produced substances best of all corresponding to the healthy organism in their physical, chemical and biological characteristics (valency, isotope condition, structural, stereoisomeric shape of the molecule, solubility, dispersity, oxidation state, half-life, safety and other parameters) (Shenderov 2014b, Bik et al. 2018, Shenderov 2011b, Sonnenburg and Backhed 2016, Wilson and Nicholson 2017). The presence in the digestive tract of various nutrients and diverse components of gut symbiotic microbiota may considerably alter the intensity of effects of probiotic bacterial metabolites or even completely eliminate their action. Violation of homeostasis of these molecules can often be a risk factor for different diseases (Shenderov 2014b, Beloborodova and Osipov 2000, Clarke et al. 2014, Maguire and Maguire 2019).

There is a considerable chemical and functional resemblance between many metabolic and signaling molecules synthesized by the cells of mammals, indigenous and probiotic microorganisms as well as micronutrients of food products. These low molecular weight compounds of dietary, endogenous and microbial

B. A. Shenderov et al., *METABIOTICS*,
https://doi.org/10.1007/978-3-030-34167-1_5

origin often act as regulators of intra- and interpopulation interaction (nervous, humoral, immune, metabolic, information and epigenetic) of all macroorganism cells, as well as between the cells of host and its symbiotic microbiota. Nondigestible or harmful substances of dietary and/or microbial origin are either digested by endogenous microbial enzymes or excreted with feces and urine. Gut microbiota is also capable of accumulating different macro- and microelements in the amounts sufficient to meet human daily demand in these elements for a period between 3 and 50 days (for example, *B*, *Mg*, *Se*, *Zn*) and even 25–190 days (for example, *Co*, *Cu*, *Mn*, *Si*) (Shenderov 2008, 2013, Nicholson et al. 2012, Shenderov and Midtvedt 2014, Sonnenburg and Backhed 2016). The metagenome of gut microorganisms is 100 times as large as the metagenome of eukaryotic cells of organs and tissues in the human gastrointestinal tract. Many biochemical metabolic pathways are absent in humans and are only provided by the genome of gut microbiota. Recent studies of the genomes of over 700 different representatives of gut microbiota have allowed to identify the genes that encode over 3200 unique chemical reactions (Bik et al. 2018). Primarily, gut microorganisms are associated with the degradation of dietary fibers, proteins and peptides, deconjugation and fermentation of non-digestible complex polysaccharides and oligosaccharides, with glycophospholipids and deconjugation and decarboxidizing of bile acids, the biosynthesis of vitamins (K and B vitamins), isoprenoids, polyphenols, cholesterol lowering, the metabolism of amino acids and xenobiotics with the production of biologically active molecules which take an active part in various human functions and reactions. For instance, daily energetic contribution of symbiotic gut microorganisms to its total energy content is comparable to the contribution of daily dietary intake through food components that are currently used for calorie calculation (Volkov et al. 2007, Panyushin 2012, Braune and Blaut 2016, Yadav et al. 2018).

In its distal part, human digestive tract is a natural bioreactor where symbiotic and commensal microorganisms actively destruct and metabolize non-digestible dietary fibers (resistant starches, cellulose, polysaccharides, oligosaccharides, pectins, unabsorbed sugars such as raffinose, lactose, stachyose etc.). Gut bacteria (*Bacteroides, Ruminococcus spp* etc.) produce a large complex of various hydrolytic enzymes that degrade the above-mentioned macromolecules. *Bacteroides*, *Bifidobacterium*, *Propionibacterium*, *Eubacterium*, *Lactobacillus*, *Clostridium*, *Roseburia*, *Prevotella* and other anaerobe microorganisms split the developing oligosaccharides into short-chain fatty acids and different gases (hydrogen, carbon dioxide, methane and others). Propionic and butyric acids can also be formed from amino acids and peptides through fermentation by the representatives of the *Bacteroides* and *Firmicutes Phyla* (Chervinets et al. 2018, Shenderov 2013b, 2015, Clarke et al. 2014, Hall and Versalovic 2018, Maguire and Maguire 2019, Shaikh and Sreeja 2017, Yadav et al. 2018). Gut microbiota takes part in degradation and modification of nitrogen-containing substances (different proteins, urea, nitrates etc.), lipids, nucleic acids, glycosides, amino sugars, chitins, organic acids, epithelial and microbial cells and other components getting into this section of the digestive tract. *Bacteroides, Clostridium, Propionibacterium, Fusobacterium, Lactobacillus, Streptococcus* and other microorganisms contained in the large intestine actively participate in the proteolysis of endogenous and exogenous proteins. Proteins of intestinal mucosa, desquamated epithelium, as well as enzymes of pancreatic and other intestinal juices are the endogenous sources of proteins needed for this purpose. Complex proteins are split by various microbial peptidases, proteases and endopeptidases. By participating in transamination, decarboxylation, deamination and dehydrogenation reactions, different gut microorganisms (*Clostridium*, *Bacteroides*, *Bifidobacterium*, *Lactobacillus*, *Peptostreptococcus* etc.) are capable of metabolizing aromatic amino acids (tyrosine, phenylacetic acid, tryptophan) along with the formation of noticeable amounts of indole, indolepropionic acid, tryptamine, kynurenine, serotonin, ammonium, pyruvate and other compounds. To date, over 85 species of indole-producing bacteria have been identified in the human gut (Hall and Versalovic 2018, Maguire and Maguire 2019, Shaikh and Sreeja 2017, Yadav et al. 2018, Zelante et al. 2013). The resulting free amino acids and short peptides are used as substrates and nutrients for subsequent catabolic processes or else undergo fermentation forming branched fatty acids (2-methyl butyrate, isobutyrate, isovalerate), organic acids, various gases (hydrogen, carbon dioxide) and small amounts of phenols, amines, indoles and ammonia (Neis et al. 2015, Shaikh and Sreeja, 2017, Yadav et al. 2018).

Gut bacteria in the human organism synthesize a broad spectrum of vitamins (niacin, biotin, riboflavin, thiamine, pantothenic acid, folate, pyridoxine, cobalamin, as well as vitamin K). B vitamins are primarily synthesized by the representatives of *Bacteroides*. It was found that riboflavin and niacin are synthesized by the representatives of 166 and 162 bacterial groups, respectively. Riboflavin and biotin are mainly synthesized by *Proteobacteria*, *Fusobacteria* and *Bacteroides species,* as well as by the *Actinobacteria* and *Firmicutes* species (Said 2013, Shaikh and Sreeja 2017, Yadav et al. 2018). Over the past few years, the research focus has been increasingly on the role of gut bacteria in polyphenol metabolism. To date, over 6000 species of various herbal phenols have already been identified, and their daily dietary intake can amount to 800 mg. Ingested polyphenol compounds belong to flavonoids, phenolic acids, stilbenes, lignans and other groups. It is important to note that the overwhelming majority of these herbal compounds reach the large intestine cavity without getting absorbed in the small intestine (Pérez-Jiménez et al. 2011, Yadav et al. 2018). Polyphenol degradation to different metabolites is observed in *Bacteroides*, *Clostridium*, *Eggerthella*, *Slakia* (Braune and Blaut 2016). When polyphenols are hydrolyzed by microbial enzymes, the resulting monomers and aglycons subsequently undergo decarboxylation and phenol ring splitting whereby hydroxyphenyl propionic acid and hydroxyphenyl acetic acid are formed. Different bacteria were noted to vary both in phenol compounds metabolism and in adsorption and excretion of intermediate and final metabolic products (Braune and Blaut 2016, Quartieri et al. 2016, Tomas-Barberan et al. 2014, Tomás-Barberán et al. 2017, Yadav et al. 2018).

Bile acids that are prerequisite for lipid metabolism are constantly being synthesized in the liver (lithocholic acid and deoxycholic acid). Gut bacteria in the digestive tract can metabolize both the side chain of these molecules and the steroid nucleus of these acids. Over thirty different bile acids circulate in the human organism. This variety is ensured by the metabolic activity of gut bacteroides, bifidobacteria, clostridia, lactobacilli and other microorganisms carrying the genes of an enzyme — hydrolase of salts of these acids. It is owing to breaking the bond between a bile acid and a conjugated amino acid that microbial detoxication of bile acids is achieved. Microbial degradation can modify bile acids through epimerization of its hydroxyl group (Clarke et al. 2014, García-Cañaveras et al. 2012, Yadav et al. 2018). Gut microbiota as well as enzymes of the host organism can degrade choline. Notably, the microbial conversion of choline results in the formation of trimethylamine, which is transformed through the action of liver enzymes into trimethylamine N-oxide whose accumulation is associated with cardiovascular diseases in humans (Clarke et al. 2014, Wang et al. 2011).

Also, digestive tract microbiota plays an important role in sulfur metabolism in the human organism. Hydrogen sulfide formation by gut bacteria is due to both bacterial enzymes participation in the fermentation of sulfur-containing amino acids (cystein, cystine, methionine, taurine, sulfur-containing mucin) and microbial (for instance, by the representatives of *Desulfovibrio*) utilization of sulfate (sulfite) ions. Presently, 60 genera and 220 species of gut microorganisms are known which can reduce sulfate-containing compounds (Shenderov 2015, Yadav et al. 2018). Discussing the digestive and other functional roles played by gut microbiota in the human organism, it should be noted that it is crucial for water and mineral homeostasis. Just enough to remind about the amount of macro- and microelements contained in the raw biomass of fecal microorganisms in adults, which exceeds their daily requirement in healthy people by 5–50 times (Shenderov 2008).

Metabolic, signaling, transport and other functions of the representatives of indigenous microbiota are more important than the quantitative content of microorganisms belonging to various species in a biotope (Oleskin et al. 2016, Shenderov 2014b, 2016, 2017, Bik et al. 2018, Blum 2017, Carding et al. 2015, Engevik and Versalovic 2017, Falony et al. 2016, Gilbert et al. 2016, Nicholson et al. 2012, Shenderov 2011b). Metabolome analysis of feces and the content of different digestive tract sections, urine, blood plasma, cerebrospinal fluid in mammals including humans, germfree and conventional mice and other experimental animals, as well as the juxtaposition of co-metabolomes of other animals and their microbiota have shown that metabolism in mammals is closely connected with their gut microbiota (Nicholson et al. 2005, Wikoff et al. 2009). Metabolome studies at the end of the previous (Midtvedt 1985, Shenderov et al. 1989, Shenderov et al. 1990, Tamm et al. 1987) and the beginning of this century (Berer et al. 2011, Clarke et al. 2014, Marcobal et al. 2013, Nicholson et al. 2012, Zierer et al. 2018, Trugnan et al. 2002), capable of detecting a great number of low molecular weight metabolites, have revealed that the concentrations of proteins, carbohydrates, peptides, short-chain fatty and other organic acids, cations, anions and other low molecular weight compounds in the feces of germfree and conventional mice and humans are essentially different (Table 1). In animals, intragastral administration of different antibiotics (lincomycin, ampicillin, azlocillin, cefalexin, tetracycline, rifampicin, sizomicin, pefloxacin) in therapeutic concentrations for 7 days led to considerable changes in the fecal excretion of amino acids, short-chain fatty acids (SCFA) and other organic

Table 1 Fecal excretion of some nitrogen and carbonic containing compounds in rats and humans (Shenderov et al. 1990)

Compound and method of its determination	Concentration or relative distribution, %		
	Rats		
	Conventional	Germfree	Humans
– total proteins, mg/g (Lowry method)	30.9 ± 7.4	80.2 ± 6.6	21.01 ± 3.8
– total aminoacid pool, μg/g (Aminoacid analyzer "Biotronik IC-5000")	708 ± 60	28,000 ± 3800	5656 ± 1800
ASP	6.6 ± 1.2	3.2 ± 0.7	9.5 ± 1.8
THR	3.6 ± 0.8	10.7 ± 2.8	7.2 ± 0.6
SER	4.2 ± 0.9	11.1 ± 2.7	5.3 ± 0.3
GLU	44.9 ± 4.3	11.8 ± 2.4	24.6 ± 2.2
ALA	8.9 ± 1.4	12.2 ± 2.7	9.9 ± 1.2
LYS	7.4 ± 2.4	5.5 ± 1.4	9.3 ± 1.0
GLY	7.2 ± 1.3	30.8 ± 2.4	4.9 ± 0.3
VAL	5.5 ± 1.6	8.8 ± 1.9	7.4 ± 0.7
MET	1.7 ± 0.7	5.8 ± 1.2	4.6 ± 0.3
ILE	6.6 ± 1.4	6.1 ± 1.9	7.1 ± 0.6
LEU	5.7 ± 1.2	8.6 ± 1.6	9.2 ± 1.4
NH_3, μg/g	114.4 ± 38.9	80.0 ± 18.0	1118 ± 700
– total carbohydrates, μg/g (reaction with phenol and H_2SO_4)	6900 ± 2200	23,300 ± 2800	12,500 ± 1500
– concentration of total SCFAs, μg/g (gas chromatography "LHM-80", USSR)	6000 ± 1400	200 ± 40	9400 ± 2500
acetic acid	85.1 ± 3.4	100	65.0 ± 2.5
propionic acid	9.0 ± 1.9	–	14.0 ± 1.3
i-butyric acid	0.7 ± 0.1	–	2.0 ± 0.2
n- butyric acid	2.7 ± 0.6	–	11.0 ± 2.1
i-valeric acid	0.8 ± 0.2	–	3.0 ± 0.5
n-valeric acid	1.0 ± 0.3	–	3.0 ± 0.6
caproic acid	–	–	2.0 ± 0.2
– the concentration of some carbonic acids, μg/g (ion exchange chromatography "Dionex" model 10, USA)			
formic acid	1500 ± 300	1800 ± 400	traces*
succinic acid	3700 ± 400	2000 ± 500	Traces
α-ketoglutaric acid	8.0 ± 1.0	–	8.5 ± 1.9
lactic acid	6.0 ± 1.0	50.0 ± 11.0	12.1 ± 1.6
pyruvic acid	150.0 ± 12.0	310.0 ± 19.0	Traces

*trace: < 2.5 μg/g

acids 5 days after antimicrobial drugs withdrawal (Table 2). At the same time, morphological changes were observed in the biofilm that covers the digestive tract mucosa; antagonistic activity with respect to opportunistic gram-negative bacteria in *in vitro* tests was decreasing, and colonization resistance of the digestive tract to externally introduced similar opportunistic pathogenic bacteria was also impaired

Table 2 Influence of antibiotics on fecal excretion (Shenderov et al. 1990)

A: of some compounds in Wistar rats (5 days after end of antibiotic intake)					
Antibiotics (mg/kg orally for 7 days)	**Ammonia nitrogen, mg%**	**% from total pool aminoacids**			
		GLU	**ALA**	**ASP**	**VAL, LEU, ILE**
Lincomycin (180)	3.0 ± 0.6	63.0 ± 5.7	8.0 ± 1.5	5.0 ± 0.9	9.0 ± 2.5
Ampicillin (180)	3.3 ± 1.1	42.0 ± 2.4	20.0 ± 3.8	6.0 ± 1.9	11.0 ± 2.8
Azlocillin (1350)	3.6 ± 0.8	19.0 ± 1.1	17.0 ± 2.9	5.0 ± 1.3	22.0 ± 4.9
Cefalexin (180)	3.0 ± 0.7	48.0 ± 2.7	12.0 ± 1.9	8.0 ± 1.7	8.0 ± 2.0
Tetracycline (90)	1.0 ± 0.2	30.0 ± 3.1	12.0 ± 2.1	2.0 ± 0.3	18.0 ± 6.3
Rifampicin (40.5)	1.5 ± 0.4	20.0 ± 4.5	14.0 ± 1.4	4.0 ± 0.4	11.0 ± 4.1
Sizomicin (12.6)	1.0 ± 0.2	31.0 ± 4.2	10.0 ± 2.5	8.0 ± 1.0	22.0 ± 7.9
Peflacin (90)	1.0 ± 0.2	34.0 ± 2.3	13.0 ± 1.8	7.0 ± 1.8	13.0 ± 3.3
Without antibiotics	3.0 ± 0.5	42.0 ± 4.5	6.2 ± 1.0	4.0 ± 0.5	8.1 ± 1.5
B: of some SCFAs in Wistar rats (5 days after end of antibiotic intake)					
Antibiotics (mg/kg orally for 7 days)	**Relative distribution (%) of SCFAs in faeces**				
	acetic acid	**propionic acid**	**n-butyric acid**	**n-valeric acid**	
Lincomycin (180)	90.0 ± 5.0	7.0 ± 3.0	3.0 ± 0.8	traces	
Ampicillin (180)	89.0 ± 6.0	6.5 ± 2.0	4.5 ± 1.0	traces	
Azlocillin (1350)	75.0 ± 5.0	20.0 ± 4.9	5.0 ± 1.7	traces	
Cefalexin (180)	83.0 ± 5.0	14.0 ± 5.3	2.2 ± 0.9	0.8 ± 0.07	
Tetracycline (90)	90.0 ± 6.0	7.2 ± 2.3	2.0 ± 0.5	0.8 ± 0.1	
Rifampicin (40.5)	87.0 ± 6.0	11.0 ± 3.0	1.6 ± 0.8	0.4 ± 0.03	
Sizomicin (12.6)	82.0 ± 7.0	15.0 ± 4.0	2.5 ± 0.7	0.9 ± 0.2	
Peflacin (90)	84.0 ± 4.0	12.0 ± 2.0	3.5 ± 0.8	0.8 ± 0.1	
Without antibiotics	84.0 ± 5.0	12.0 ± 3.0	3.0 ± 1.0	1.0 ± 0.5	
C: of some carbonic acids in Wistar rats (5 days after end of antibiotic intake)					
Antibiotics (mg/kg orally for 7 days)	**Ratio:**				
	Pyruvic acid	**Lactic acid**	**α-ketoglutaric acid**		
Lincomycin (180)	10	18	23		
Ampicillin (180)	10	35	26		
Azlocillin (1350)	10	32	30		
Cefalexin (180)	10	34	19		
Tetracycline (90)	10	2	20		
Rifamicin (40.5)	10	35	20		
Sizomicin (12.6)	10	33	27		
Peflacin (90)	10	41	34		
Ampicillin (180)	10	35	26		
Without antibiotics	10	0.5	0.3		

(Shenderov et al. 1990). Feeding mice with foodstuff containing particular components has helped modify the spectrum and quantitative content of metabolites in animal urine and feces; it has led the researchers to conclude: diet plays a crucial role in changing gut microbiota and host metabolism functionality (Marcobal et al. 2013).

In recent studies of gut microbiome in 786 persons, mono- and dizygotic twins aged 55 to 65 and predominantly female (Zierer et al. 2018), over 1116 metabolites (amino acids, nucleotides, sugars, vitamins, fatty acids, other compounds and products of their intermediate metabolism) have been identified in blood and fecal samples. Among them, 469 similar metabolites have been simultaneously detected both in fecal and blood samples; 647 metabolites were unique for the feces of study participants. Over 36% of all compounds detected in blood were microbial in origin (Hood 2012); for example, 1 to 20% of lysine and threonine contained in blood plasma of an adult is synthesized by gut microbiota. The microbial origin of 40% of metabolites in human blood was also demonstrated by other researchers (Beloborodova and Osipov 2000). It was found that the peak of molecular microbial markers in blood serum is proportional to the biomass of specific microorganisms. In light of this, Russian researchers (Beloborodova and Osipov 2000) offered a theory of homeostasis of low molecular weight microbial molecules which was for the time innovative. According to the theory, their content in biological fluids is the most important regulatory mechanism of symbiotic relationship between the host and his microbiota; violation of homeostasis of these molecules is a risk factor for various diseases. They can boost or inhibit the activity of different microorganism cells, or be indifferent to it. In a recent publication, it has been suggested that "overfeeding" leads to an increased activity and changed functionality of the microbiota, thus severely disturbing host-microbe interactions and leading to dysbiosis and disease development (Lachnit et al. 2019).

Low molecular weight microbial compounds can act as metabolic molecules, precursors of co-factors of bioactive compounds, signaling molecules and molecules at the same time having metabolic and signaling activity. Their effects can be displayed on the molecular level (DNA and chromatin structure; RNA interference; post-translational modification of gene products) and on the cell level (on cell surfaces and membranes, in mitochondria and ribosomes), inside cell cytoplasm, in intercellular space, in tissues, organs, physiological systems, as well as on the level of the whole organism (Carding et al. 2015, Shenderov 2011a). By origin, most metabolites found in human biological fluids are not related to the metabolic activity of its own cells; symbiotic microbiota is largely responsible for human metabolome composition. Until now, the total number of potential gut bacteria metabolites remains impossible to determine; they include molecules of dietary, endogenous and microbial origin. One can only speculate that there are thousands and millions of metabolites of microbial origin or those produced as a result of microbial transformation of substrates and metabolites of different nature. In different persons, many microbial metabolites are structurally and functionally similar; but some of them are unique and can be found only in a few individuals. In any case, when estimating human metabolome, it should be borne in mind that microbial metabolites, as well as those microbially transformable, can act both as health promoting factors and as agents involved in disease processes (Dorrestein et al. 2014, Trugnan et al. 2002). It should be noted that the molecular mechanisms through which symbiotic microorganisms influence different physiological functions, biochemical and behavior reactions of the human organism, as well as how they interact among themselves and with host cells, — all this so far remains largely unclear.

Factors and Agents that Modify the Composition and Functions of Symbiotic Microbiota; Diagnostic Methods for Microecological Imbalance and its Consequences

According to the latest data, human microbiome variability is only by 10% related to individual genetic traits; microbiome differences between individuals are largely associated with the effects of various endogenous and exogenous factors: diet in the first place (Blum 2017; Falony et al. 2016; Zhernakova et al. 2016). Out of 69 evaluated factors, various therapeutics (over 10% varieties) are playing the most important role in modification of gut microbiota composition (Falony et al. 2016). Starvation, low physical activity, diets with increased levels of sugar, fat, with low dietary fiber content, processing aids and heavy metal salts contained in foods, alcohol consumption, pesticide and radiation exposure, the influence of space flights, surgeries, bacterial and viral infections, other factors and agents or their combinations can reversibly or irreversibly alter human microbial ecology (Shenderov 2014b; Carding et al. 2015; Dietert and Dietert 2015; Maguire and Maguire 2019; Pflughoeft and Versalovic 2012; Shenderov 2011b; Sonnenburg and Backhed 2016). Of all pharmaceuticals, antibiotics have the most pronounced negative effect on human indigenous microbiota. Many immunosuppressors, antihistamines in pharmacological concentrations also inhibit the growth of bifidobacteria, lactobacilli, enterococci, Escherichia coli and other commensal and symbiotic gut microorganisms. Also, microecological disorders are caused by the administration of local anesthetics, absorbents, nauseants, enveloping agents, laxatives, expectorants, choleretics and other therapeutics. Certain colorants, nitrites, nitrates and some hormones can be potential dysbiotic agents (Shenderov 2014b; Maguire and Maguire 2019). According to Swedish researchers' data (Bengmark 2013), half of the 2000 pharmaceuticals registered in this country can cause side effects in the human digestive tract (nausea, vomiting, diarrhea, constipation etc.) associated with microbiota imbalance in this system. In a context where the compensation abilities of the host-microbiota system are exceeded by negative effects on microbial ecology in length and intensity, microecological disorders (dysbioses) develop, as well as the imbalance of systems controlling intra- and interpopulation symbiotic relationship between the host and its microbiota and, consequently, the risks of numerous diseases. Thus, the negative stress effects of a multitude of biogenous and abiotic

B. A. Shenderov et al., *METABIOTICS*,
https://doi.org/10.1007/978-3-030-34167-1_6

factors conflict with adaptive capabilities of modern humans and lead to a considerable unbalancing of those gut microbiota functions that are connected with maintaining dietary, metabolic, epigenetic, neurohormonal and immune homeostasis (Shenderov 2008; Shenderov 2016a).

According to contemporary views, chronic diseases and premature ageing in humans are associated with the imbalance of energy metabolism and redox processes in tissues, accelerated cell ageing, imbalance of cell proliferation and apoptosis, diminishing of stem cell pool, chronic inflammation, loss of proteostasis, mitochondrial disfunction, telomere shortening, microecological imbalance in gastrointestinal tract, nutritional disorders and violation of intrapopulation and intercellular relations between eukaryotic and prokaryotic cells, genome instability, epigenetic changes in gene expression, modification of information signaling pathways on the cell surfaces and membranes and inside the cells (Shenderov 2013c; Shenderov 2016a; Dietert and Dietert 2015; Kanherkar et al. 2014; López-Otín et al. 2013; Suvorov 2013). The degree of manifestation of these cellular and molecular parameters, as well as a relative "balance" of pathophysiological processes are responsible for the phenotypic manifestation of ageing, of a definite pathological syndrome or a particular chronic disease (Shenderov 2008; Shenderov 2016a; Bomba et al. 2012; Hanahan and Weinberg 2011; López-Otín et al. 2013; Shenderov and Midtvedt 2014; Sonnenburg and Backhed 2016).

A profound microecological imbalance of natural microbiocenoses in all spheres of our planet is the key factor of health deterioration in the majority of its inhabitants. The question arises of preserving not only human life on the Earth, but the whole existing variety of living organisms (Shenderov 2008). Profound and long-term changes in the structure and quantitative content of human indigenous microbiota often induce the risk of gastrointestinal diseases (diarrhea, constipation, colitis, irritable bowel syndrome, gastritis, gastric ulcer, other chronic diseases), neurodegenerative diseases (Alzheimer's disease, Parkinson's disease, Huntington's disease, Amyotrophic lateral sclerosis, Friedreich's ataxia), metabolic syndrome (atherosclerosis, Diabetes mellitus type 2, obesity, gout), autoimmune disorders (multiple sclerosis, Diabetes mellitus type 1, systemic lupus erythematosus), behavior and mental disorders (autism, schizophrenia, chronic fatigue syndrome), musculoskeletal disorders (fibromyalgia, skeletal muscle hypertrophy/atrophy, nonspecific arthritis and arthrosis), kidney stone disease and gallstone disease, bronchial asthma, atopic dermatitis, psoriasis, neoplasms, menstrual disorders, infertility, preterm birth, neonatal anemia, cachexia, transplant against host syndrome, opportunistic endo- and superinfections of different localizations (Shenderov 2014b; Shenderov 2016a; Carding et al. 2015; Frank et al. 2011; Gilbert et al. 2016; Maguire and Maguire 2019; Sonnenburg and Backhed 2016). As any given manifestations of microecological imbalance are currently found in a considerable number of people living in countries with developed economies, it becomes clear how important it is to stop further destruction of human microbial ecology and living environment. It explains the need for still greater attention of a wide range of experts to the research into normal and pathological human microbial ecology as the primary target for many physical, chemical and biological agents. Along with this, the available data

Table 1 Methods and technologies used for the characteristics of human microbiota components and functions

Item No.	Methods and technologies
1	Different kinds of microscopy
2	Classic bacteriological method
3	Molecular and genetic methods
4	Different kinds of chromatography and chromatography–mass spectrometry
5	Application of specific biochips
6	Use of germfree animals and different gnotobiotes
7	OMIC-technologies

allows us to think that maintaining and restoring normal human microbiota should be considered as one of the most vital scientific experimental problems for health preservation both in individuals and in the whole human population.

Assessment of symbiotic microbiota and detection of violations in its structure and functions under exposure to various stress factors and agents, as well as monitoring of their restoration in case of prescription of microecological agents, is currently done by (Bezrodny and Shenderov 2016; Shenderov 2012; Shenderov 2015; Berer et al. 2011; Gilbert et al. 2016; Karczewski and Snyder 2018; Meisel and Grice 2017; O'Flaherty and Kleenhammer 2011; Sonnenburg and Backhed 2016) both traditional tests and post-genomic OMIC-technologies (Table 1). The best-known among the latest techniques are the methods based on the research into molecular genomics, transcriptomics, proteomics, metabolomics, epigenomics, phenomics and bioinformatics.

Contemporary Microecological Strategies of Gut Microbiota Modulation for Human Health Preservation, Restoration and Improvement

For health preservation and reducing the risk of premature ageing and related metabolic diseases, human diet should include various traditional, organic and functional foods as well as therapeutics that support alimentary and other functions of gut microbiota. To create or restore the impaired human biocenoses, various targeted microecological techniques and methods are used (Table 1), with a focus on either prevention or treatment. In the longer term, these approaches can be used in microecological engineering of pregnant and breastfeeding mothers, as well as infants for shaping their targeted microbial ecology. It should be remembered that though the responses of the host-microbiota system to the implementation of different dietary and microecological agents are on the whole predictable, they can vary considerably from individual to individual as far as their focus, efficiency and distinct manifestations are concerned (Shenderov 2014b; Lenoir-Wijnkoop et al. 2007; Shenderov 2011b; Sonnenburg and Backhed 2016). Since 1950s, over 150 different microecological therapeutics have been developed and produced commercially for prevention and treatment of diseases related to the imbalance of symbiotic microbiota. To preserve and restore human microbial ecology, a wide range of microecological therapeutics are used (probiotics, symbiotics, combiotics, prebiotics, synbiotics, virobiotics (including phagobiotics), genetically engineered probiotics, metabiotics), as well as the technique of transplantation of large intestine microbiota.

The term "probiotics"is of Greek origin and means "for life" and was introduced into the scientific literature in the 50s of the last century by the German nutritionist Werner Kollath in contrast to the word "antibiotics". Most foreign experts in this field understand the term "probiotics" as living microorganisms that, when administered in an adequate amount, have a positive effect on the health of the host (Reid et al. 2011). In Russia, this term often refers to living microorganisms or substances of any origin, which in the natural way of appointment can have beneficial effects on physiological functions, metabolic and behavioral reactions of the body by optimizing its microecological status (Shenderov 2001; Shenderov 2008; Shenderov 2011b). Symbiotics are probiotics that include two or more strains of live probiotic microorganisms that provide additional, more often stimulating effects (a typical

B. A. Shenderov et al., *METABIOTICS*,
https://doi.org/10.1007/978-3-030-34167-1_7

Table 1 Principles and techniques for correction of the digestive tract microbial ecology

Item no.	Principles and techniques
1	The use of functional foods including those containing poly- and oligosaccharides of plant and microbial origin designed to the restoration of human nutritive and microbial statuses.
2	Optimization of pH and redox potential in the digestive tract by prescribing sorbents, antioxidants, detoxication agents etc.
3	Prescription of immunostimulants of different genesis that enhance secretory immunoglobulin production and improve other mechanisms of local and systemic immunity
4	Injections of anti-adhesive antibodies and lectins that block the adhesive capability of potential pathogenic microorganisms
5	Selective decontamination of the digestive tract through the use of specially chosen non-absorbing antibiotics, bacteriophages, antimicrobial peptides and other compounds
6	Optimization of pregnant and breastfeeding women microbiota, newborns inoculation with certain microorganisms and their complexes, prescription to children and adults of probiotics, prebiotics, symbiotics, synbiotics on the base of the representatives of normal gut microbiota contained in pharmacopeial drugs, dietary supplements or special-purpose food products
7	The application of therapeutics contributing to selective reproduction of "human-friendly" anaerobe representatives of symbiotic microbiota
8	Enrichment of food products with dietary supplements, structural components, metabolites and signaling molecules isolated from known strains of probiotic microorganisms

example of the Russian symbiotic is "Bifikol", which includes E. coli, removing excess oxygen from the intestinal lumen, and bifidobacteria, which in these conditions multiply more intensively, producing increased amounts of biologically active substances (for example, vitamins that stimulate the growth of intestinal microorganisms). Prebiotics - a variety of non-digestible carbon-containing compounds, mainly plant polysaccharides and oligosaccharides, which have a positive effect on the host, selectively stimulating the growth and activity of representatives of the intestinal microbiota, relating primarily to lactobacilli and bifidobacteria. Synbiotics are probiotic complex products containing live probiotic microorganisms and prebiotic substances that stimulate their growth. Combiotic are variants of synbiotic agents, which in addition to probiotic microorganisms and prebiotics, added other functional food ingredients (eg, vitamin and mineral premixes, phenolic or other plant compounds, etc.). To date, the demand has been the greatest for prebiotics and synbiotics highly diverse in composition (Tables 2 and 3) (Chervinets et al. 2018; Shenderov 2001; Shenderov 2014b; Dore et al. 2013; Lenoir-Wijnkoop et al. 2007; Maguire and Maguire 2019; Mimee et al. 2016; Reid et al. 2011; Roberfroid et al. 2010; Shenderov 2011b; Shenderov 2013a; Sonnenburg and Backhed 2016; Suvorov 2013; Venema and Carmo 2015; Vyas and Ranganathan 2012).

Probiotic effectiveness of microorganisms is determined by their adaptive and probiotic potentials (Lebeer et al. 2008; Lebeer et al. 2018; Shenderov 2011b). Adaptive potential includes the resistance of probiotic bacteria to physical and

Table 2 The list of certain probiotics commercially available in Russia

Name	Composition
Acilact	Three strains of Lactobacillus acidophilus
Bactisubtil, Sporobacterin	strains of bacilli (*B. cereus*)
Bilactin	strains of *Enterococcus faecium* M-3185 and *E. faecium* M
Biovestin, Biovestin-lacto	strains of *B. bifidum*, *B. adolescentis*, *L. plantarum*
Biosporin	strains of *B. subtilis* and *B. licheniformis*
Bifidin	strain of *B. adolescentis* MC
Bifidumbacterin forte	strain of *B. bifidum* N1, sorbed by coal
Bifilin	strain of *B. adolescentis*
Bifilong	strains of *B. bifidum* and *B. longum*
Bificol	strains of *E. coli* and *B. bifidum*
Bifiform	strains of bifidobacteria (*B. longum*) and enterococci (*E. faecium*)
Gastropharm	strain of *L. bulgaricus* LB-51
Colibacterin	strain of *E. coli* M-17
Lactobacterin	strain of *L. plantarum* 8PA-3
Linex	strains of *L. acidophilus*, *B. infantis* and *E. faecalis*
Normoflorin L and B	strain of lactobacilli or bifidobacteria
Polybacterin	strains of *B. bifidum* ЛВА-3, *B. longum* B379M, *B. breve* 79–119, *B. adolescentis* ГО-13, *L. acidophilus* NK-1, *L. plantarum* 8PA-3, *L. fermentum* 90-T-4
Florin forte	sorbed by activated charcoal *B. bifidum* 1 and *L. plantarum* 8PA-3
Enterol	yeast cells (*Saccharomyces boulardii*)

chemical stresses (pH value, oxidative and osmotic stresses), the activity of adaptive metabolism (the capability of utilization of carbohydrate and other substrates), as well as the capability for adhesion and its activity (the amount and type of surface mucin- and fibronectin-binding proteins, exopolysaccharides, lipoteichoic acids etc.). The probiotic potential is connected with the production by live microorganisms of a multitude of low molecular weight compounds different in chemical origin (autoinducers, chemokines, modulins, effectors, substrates, co-substrates, enzymes, co-factors, metabolites, signaling molecules), similar to those of endogenous or dietary origin, as well as microbial compounds eliminating (or suppressing) the growth of pathogenic and opportunistic microorganisms in the digestive tract of the user of these therapeutics and/or positively interacting with his autochthonous microbiota and gut epithelium, improving local and systemic immunity, activating metabolic and other processes, localized or taking place outside the gastrointestinal tract (Shenderov et al. 2018a; Bik et al. 2018; Engevik and Versalovic 2017; Shenderov 2011b). Recent studies of the metabolomic genome of over 700 different representatives of gut microbiota allowed for identifying the genes encoding over 3200 unique chemical reactions (Bik et al. 2018).

Table 3 The list of prebiotics, synbiotics and combiotics commercially available in Russia

Name	Composition
Astrolin	80% of purified inulin isolated from Jerusalem artichoke tubers
Bilaminolact	*E. faecium* L-3 + bifidobacteria + plant extracts
Laminolact	*E. faecium* L-3 in combination with plant extracts
Bioflor	*E. coli* M-17 + extracts of soybeans, vegetables and propolis
Bifilis	bifidobacteria (*B. bifidum*) + lysozyme
Bifistim	*B. bifidum, B. breve, B. infantis, B. longum, B. adolescentis* + vitamins (B1, B2, B3; B5, B6, B12, H, C, E, A, D3, folic acid) + cellulose, pectin + *Ca* hydrophosphate + fructose
Vitaflor	lactobacilli, vitamin C, baker's yeast autolyzate, β-carotene
Duphalac, Lactusan, lactulose	disaccharide consisting of fructose and galactose
Normospectrum (multispectrum)	*B. bifidum* 1 and 791, *B. longum* B379M and J-3, *B. adolescentis* GO-13, *L. plantarum* 8PA-3, *L. acidophilus* NK-1 and K3Ш24, *L. casei* KHM-12, vitamins (B1, B2, B3; B5, B6, B12, H, C, E, folic acid), minerals (*Zn, Se*), inulin
Primadophilus	*B. infantis, B. longum, L. casei subsp. rhamnosus* and maltodextrin
Rekicen-RD, Fervital, Eubicor	wheat of rye bran with wine yeast sorbed on them
Stimbifid	inulin, oligofructose, vitamin and mineral premix (*Zn, Se*, vitamins B1, B2, B3, B5, B6, B12, H, C, E, folic acid)

With the growth of the market of therapeutics that improve microbial ecology, new approaches to their construction and use were being developed and known methods were being modified. The mode of action of traditional probiotics and synbiotics was being detailed, additional targets for known and new probiotics were being located (the capability of antioxidative, anti-inflammatory, anti-mutagenic activity, the ability to influence mental status, epigenotype, quorum-sensing regulation etc.). Probiotic strains were being selected not only among traditional symbiotic bacteria (Lactobacilli, Bifidobacteria, Enterococci, Escherichia, Bacilli), but also among anaerobe kinds of bacteria (for example, Bacteroides) and the kind of microorganisms that are always present, if sparingly, in one biotope or another. Attempts were being made to create symbiotic probiotics on the base of strains simultaneously participating in the realization of a number of particular physiological functions, biochemical and behavioral reactions; it was attempted to enrich the composition of probiotics and synbiotics with nanoparticles containing certain low molecular weight compounds or their groups ("functional" probiotics).

Probiotics are differentiated into those intended for the majority of the human population or for specific cohorts including autoprobiotics for personal use. Moreover, probiotics on the base of genetically engineered strains have been worked out by artificial introduction into their genome of genetic determinants responsible for the synthesis of bacteriocins, antimicrobial enzymes, immunoglobulins, interferons and other compounds that enhance adaptive and/or probiotic potential of probiotic microorganisms (Shenderov 2014b; Browne et al. 2016; Maguire and Maguire 2019; Mimee et al. 2016; Sonnenburg and Backhed 2016).

Probiotics on the base of live symbiotic bacteria have quite a long history of consumption; but until now it remains unclear why a few hundred milligrams or even grams of probiotic bacteria can cause positive changes in physiological functions, biochemical and signaling reactions and in human health on the whole, when an adult's large intestine alone contains between 1.5– 3 kg of live microorganisms of hundreds and even thousands of different species (Bengmark 2001). In the same way, it is difficult to answer the question why live probiotic microorganisms, which are foreign for people consuming them for a long time, do not cause an immediate rejection immune response or allergic reactions, as is often the case with respect to dozens and hundreds of billions of other foreign microorganisms that are daily fed into any person's digestive tract with food, water, inhaled air or as a result of contact with live and inanimate objects of the environment.

Molecular Language of Symbiotic Gut Microorganisms

In the 1990s, it was found that various microorganisms, their cell fragments, colloid particles, food starch granules and other compounds similar in size to bacterial cells or smaller can easily penetrate intestinal mucosa. Owing to contemporary chromatography and mass spectrometry methods, many similar low molecular weight particles and molecules are easily detectable in blood, urine and other human biological fluids. Their appearance in these fluids can be detected as early as a few minutes after oral administration of any given indicator compounds or microorganisms (Freter and Nader de Macias 1995; Osipov and Verkhovtseva 2011). It was shown clinically and experimentally that effects of probiotics on human and animal organisms arise in case of administration not only of live but also of killed (by heating, radiation etc.) microorganisms, as well as their cell-free extracts, purified cell walls, supernatants of culture fluids for bifidobacteria and lactobacilli (Lee et al. 2002; Reading and Sperandio 2006). The results of such studies have convincingly demonstrated that the translocation of probiotic bacteria, their cell fragments, metabolites and signaling molecules, as well as low molecular weight compounds formed as a result of microbial degradation of epithelial cells, saliva, intestinal juices, food substrates from the intestinal lumen is a common physiological process. Biologically active compounds (autoinducers) associated with the metabolic activity of symbiotic (probiotic) microorganisms can potentially modify and take part practically in any physiological function, metabolic, signaling, behavioral reaction, in intra- and intercellular information exchange. Given that, the mechanisms of their effects may be different in different symbiotic and probiotic microorganisms. Autoinducers (chemokines, modulins) of microbial nature allow symbiotic microorganisms to discern the environment, interact with each other and with host cells. These molecules trigger a cascade of processes in prokaryotic and eukaryotic cells, when their quantity reaches a certain level (*"quorum"*). Autoinducers interact with cell receptors, with regulatory proteins recognizing them and, in the long run, activate (induce) the work (expression) of the corresponding genes in the DNA of microbial, mitochondrial and eukaryotic host cells. Thanks to autoinducers, microorganisms and cells of macroorganism exchange the information and coordinate their activity. That

B. A. Shenderov et al., *METABIOTICS*,
https://doi.org/10.1007/978-3-030-34167-1_8

is why such signaling molecules are viewed in scientific literature as "words" in the molecular information "language".

Numerous research in the past two decades has shown that symbiotic (probiotic) microorganisms synthesize and discern a broad range of autoinducers of diverse chemical nature (Vakhitov et al. 2005; Shenderov 2008; Shenderov 2009; Carding et al. 2015; Engevik and Versalovic 2017; Lebeer et al. 2018; Schauder and Bassler 2001; Shenderov 2013a; Taga and Bassler 2003). The best known among them are volatile and other organic acids, lactones, peptide pheromones, furanones and other autoinducers involved in realization of the quorum-sensing phenomenon, proteins, adenosine triphosphate and other compounds produced under stress, various proteins, peptides and amino acids, various simplest metabolites of microbial cells (CH_4, H_2S, NO, CO, H_2, H_2O_2 etc.), nucleic acids, nucleotides, nucleosides, vitamins (mostly B vitamins, biotin, folic and pantothenic acids, vitamin K), short- and long-chain fatty acids, amino acids, amines, polyamines, hormone-like substances, neurotransmitters, regulatory molecules of different chemical nature involved in the *quorum sensing* signaling, polysaccharides, oligosaccharides, various surface proteins (pili, fimbriae, flagella etc.), mucins, peptidoglycans, lipoteichoic acids, numerous bioactive peptides, glycopeptides, lipopolysaccharides, antimicrobial compounds of different chemical structure, lectins, biosurfactants, pigments etc. For example, it is known that the representatives of commensal and symbiotic bacteria produce over two dozens of various antimicrobial substances alone (lactic, acetic, butyric, benzoic and other organic acids, hydrogen peroxide, carbon dioxide, nitrogen oxide, diacetyl, bacteriocins, microcins, antibiotics, defensin-like peptides, lysozyme, biosurfactants, lectins etc.) (Chervinets et al. 2018; Shenderov 2017; Shenderov and Lakhtin 2004; Shenderov et al. 2018a; Aguilar-Toaláa et al. 2018; Clarke et al. 2014; Engevik and Versalovic 2017; Gu and Li 2016; Lebeer et al. 2018; Oleskin et al. 2017; Shaikh and Sreeja 2017; Shenderov 2011a; Shenderov 2013a; Singh et al. 2018). It should be noted that molecular mechanisms through which symbiotic microorganisms influence various physiological functions, biochemical and behavioral reactions of the human organism, as well as how they interact with each other and host cells — all this remains until now in many respects unclear. The main "energetically significant" metabolites of symbiotic microbiota in the digestive tract are represented by fatty acids (lactate, acetate, propionate, butyrate, succinate), alcohols and gases (hydrogen, methane) (Panyushin 2012; Aguilar-Toaláa et al. 2018; Shaikh and Sreeja 2017; Singh et al. 2018). Apart from energy synthesis, in the process of complex microbial transformation including oxidation, reduction, the formation of chelate complexes, various stereoisomeric forms, isotope fractions, other physical and chemical modifications, lots of low molecular weight compounds are formed from the above dietary and endogenous substrates in the best available form for host cells, which are then absorbed and incorporated in enzymes and other physiologically active substances.

Microorganisms of the digestive tract are the most important source of microbial metabolites, as they secrete various terminal metabolites and/or their intermediate products throughout their life (Carding et al. 2015; Dorrestein et al. 2014; Engevik and Versalovic 2017; Gilbert et al. 2016; Nicholson et al. 2012; Singh et al. 2018;

Lyu and Hsu 2018). Some microbial metabolites are the result of work of the genes obtained by microorganisms in course of horizontal or vertical transfer of genetic material which is ever-present in microbial populations of the digestive tract; others are formed as a consequence of microbial transformation of substrates formed by host cells. Microbial transformation of exogenous substances ingested from the environment (pharmaceuticals, other organic compounds including food colorants, thickeners, many xenobiotics of household chemicals) can also be accompanied by the formation of products of their microbial modification with their entering intestinal lumen. By now, hundreds and even thousands of low molecular weight compounds had been identified that are formed as a result of metabolic activity of gut microbiota or being low molecular components of microbial cells. Many compounds of microbial origin are present in healthy and ill people, both in intestinal lumen and in biological fluids outside it, as substrates, co-substrates, enzymes or co-factors of metabolic reactions in the macroorganism, that is, molecules involved in the regulation of intra- and interpopulation interactions between prokaryotic and eukaryotic cells, as well as between the host and its microbiota. They may activate, inhibit or be indifferent with respect to various prokaryotic and eukaryotic host cells (Shenderov 2013c; Shenderov 2017; Shenderov et al. 2018a; Gilbert et al. 2016; Nicholson et al. 2012; Shenderov 2011a; Sonnenburg and Backhed 2016). Metabolome studies of intestinal content, liver, blood plasma, urine, fecal extracts have shown that any given effects of symbiotic (probiotic) microorganisms on a live organism are related to a mixture of various microbial compounds rather than one low molecular compound in particular (Lebeer et al. 2018; Martin et al. 2008; Sonnenburg and Backhed 2016). In this respect, the chemical dialogue between symbiont microbes and host cells includes both signal exchange via low molecular metabolites, proteins and peptides, and indirectly through neurohormonal, immune and epigenetic mechanisms (El Aidy et al. 2016; Maguire and Maguire 2019; Shenderov 2016b). Grown on milk base or soymilk, germfree filtrates of various probiotic strains of lactobacilli, bifidobacteria and other symbiotic microorganisms contained peptides, glycoproteins, enzymes, exopolysaccharides, peptidoglycans, SCFA and other low molecular metabolites that inactivated different mutagens, carcinogens and prevented metastatic spreading of tumor cells (Aguilar-Toaláa et al. 2018; Dzutsev et al. 2017; Shaikh and Sreeja 2017; Sharma and Shukla 2016).

Microbial molecules interact with surface, membrane, cytoplasmic, mitochondrial and nuclear receptors of epithelial cells, causing induction of different genes, maintaining genome and microgenome stability, modulating epigenome program of development and realization, ensuring intra- and intercellular information exchange and interaction between microorganisms and the host. Bioactive molecules synthesized by the representatives of indigenous microbiota lead to both local and systemic effects through neuroendocrine, immune, metabolic and epigenetic mechanisms. Many microbial metabolites influence metabolism by interacting with specific receptors on host cell membranes and nuclear structures (Oleskin and Shenderov 2016; Shenderov 2008; Bourdichon et al. 2012; Cani 2018; Carding et al. 2015; Clarke et al. 2014; Engevik and Versalovic 2017; Galland 2014; Gilbert et al. 2016; Nicholson et al. 2012; Paul et al. 2015; Sampson and Mazmanitan 2015;

Holmes et al. 2011; Lebeer et al. 2018; Shenderov 2011a; Shenderov and Midtvedt 2014; Stilling et al. 2014; Singh et al. 2018). As previously stated, digestive tract microorganisms actively metabolize complex substrates ingested with food or endogenous in origin (components of saliva, digestive juices, desquamated epithelium, dead microorganisms etc.). Microbial transformation of various substrates results in the formation of many hundreds (possibly, thousands) of low molecular biologically active functional ingredients (Shenderov 2009; Carding et al. 2015; Chilloux et al. 2016; Corthesy et al. 2007; Engevik and Versalovic 2017; Shenderov 2011b; Sonnenburg and Backhed 2016). In accordance with the latest data (Aguilar-Toaláa et al. 2018; Clarke et al. 2014; Lebeer et al. 2018; Lyte 2013; Maguire and Maguire 2019; Lyu and Hsu 2018; Oleskin et al. 2017; Singh et al. 2018), gut microbiota is proposed to be viewed as the most important structural and functional component of a single neuroendocrine and immune organ. Candidate gut microbiota hormones are SCFA (acetate, butyrate, propionate), neurotransmitters (serotonin, dopamine, noradrenaline, GABA), precursors of neuroactive compounds (tryptophan, kynurenine, L-DOPA), secondary bile acids, host metabolites (trimethylamine), cortisol, digestive tract hormones (ghrelin, leptin, glucagon-like peptide 1, PYY).

To better understand the importance of this complex function of gut microbiota in the physiology and pathophysiology of hormonal-neuro-immune axis of the human organism, let us have a closer look at its work. In the organisms of mammals, apart from specialized hormone-producing glands (the adrenal glands, thyroid gland, hypophysis etc.), the brain, the spinal cord and immune organs (the thymus, lymph gland, bone marrow etc.), diffuse regulatory endocrine system (APUD-system) is also found in many organs and tissues. The majority of its cells are present in the gastrointestinal tract, producing various signal agents (hormones) that regulate digestive juice secretion, appetite, mood, vascular tone and other mammal functions (Yaglov and Yaglova 2012; Demeure 1993; O'Callaghan et al. 2016). By now, over 20 gastrointestinal hormones have been identified (bombesin, gastrin, histamine, ghrelin, glucagon, galanin, catecholamine, leptin, motilin, melatonin, peptide YY, secretin, serotonin, somatostatin, substance P, beta-endorphin, enkephalin, enteroglucagon etc.) that are formed by diffusely distributed cells (APUD cells) of the APUD system localized throughout the whole digestive tract length. Many of the previously specified tissue mediators and hormones are of peptide nature, they are used locally and are also fed to systemic blood circulation; as regards their chemical composition and functions, they often bear similarity to brain neurotransmitters and compounds produced by gut microbiota representatives (Oleskin et al. 2017; Singh et al. 2018; Tan et al. 2015; Tsang et al. 2013).

Also, gut microorganisms play a fundamental role in the functioning of mammal immune system. When the performance of the "symbiotic microbiota – immunity" system is optimal, humans can successfully resist pathogenic microorganisms and support the work of the regulatory mechanisms that ensure tolerance to harmless antigens. Microbial metabolites and microbial cell components produced in the gut are necessary for keeping immunologic balance (Aguilar-Toaláa et al. 2018; Engevik and Versalovic 2017; Forbes et al. 2016; Laino et al. 2016; Oleskin et al. 2017).

Among the metabolites of symbiotic bacteria having effect on immune composition in mammals, the best studied are SCFA formed in course of microbial fermentation of undigestible dietary fibers (Shenderov 2013b; Engevik and Versalovic 2017). Microbial SCFA modulate immune responses through the activation of chemoattraction receptors GPR41 and GPR43, localized in different immune cells, and/or through inhibiting the activity of intracellular histone deacetylases. Depending on the type and concentration, SCFA can decrease neutrophil adhesion and chemotaxis, suppress immune cell infiltration from blood stream to the inflammation site. Also, these acids regulate the production of a range of leukocytic cytokines (TNF-α, IL-2, IL-6, IL-10), eicosanoids and chemokines (MCP-1, CINC-2). Acetate and butyrate modify the activity of neutrophils and macrophages that get involved in inflammation processes through the induction in epithelial and immune killer cells of the corresponding cytokines that take part in cell migration, leukocyte chemotaxis and suppression of adhesion molecule formation. These SCFA inhibit the activity of sirtuins and NF-κB involved in attenuation of immune response to LPS of gram-negative bacteria (Beloborodova 2012; Golovenko et al. 2011; Shenderov 2013b; Canani et al. 2011), suppress T-cell proliferation and activation (Meijer et al. 2010), reduce antibody synthesis and content in peripheral blood (Erofeev et al. 2012; Faith et al. 2014), cause apoptosis in lymphocytes, macrophages and neutrophils (Shenderov 2013b).

It is assumed that some SCFA (butyrate, propionate, but not acetate) and other metabolic products of symbiotic microorganisms can modify the balance between pro- and anti-inflammatory immune mechanisms by enhancing the egress and activity of extrathymic (intestinal) anti-inflammatory regulatory T cells (Tregs) and by affecting dendritic cells (DC) involved in the process of their formation. SCFA effect on Tregs is attributed to the change in epigenetic gene regulation in immune cells caused by these organic acids (Faith et al. 2014; Shaikh and Sreeja 2017). Detailed research into the mechanisms of microbial activation of the immune system has shown that immune response modulation is achieved by not only SCFA produced by host microbiota cells, but also by their various components (skeletons, bacterial liposaccharides, peptidoglycans, surface polysaccharides and proteins, nucleic acids, peptides and their complexes). Surface polysaccharide and protein structures of probiotic lactobacilli elicit anti-inflammatory TNF-α immune response in peripheral blood mononuclear cells, while anti-inflammatory immune effects are caused by the joint action of microbial metabolites and surface components of these bacteria. LPS of *E. coli* is by 100 to 1000 times as effective in stimulating the synthesis of cytokines (TNF-α, IL-6, IL-10) by peripheral blood monocytes as peptidoglycans; the latter, in turn, were 10 to 100 times more effective than microbial lipoteichoic acid.

Metabolites and surface structures (cell walls, peptidoglycans, surface polysaccharides) of probiotic bacteria reveal their anti-inflammatory effect through increased induction of IL-10 synthesis along with other cytokines (Ashraf et al. 2014; Laino et al. 2016). The representatives of gut commensal and symbiotic microbiota (clostridia, bacteroides, bifidobacteria, propionibacteria, lactobacilli etc.), their components and metabolites interact with human immune system via ligand-receptor link with Toll- and NOD-like cell receptor families. The crucial role belongs to TLR receptors; they are localized in macrophages, neutrophils, dendritic cells, endothelial and other cells. TLR1, TLR2, TLR6 recognize microbial lipoproteins, TLR4 — lipopolysaccharides (LPS), TLR5 — flagellin, TLR7, TLR8 — single-chain RNAs, TLR9 — non-methylated DNA. TLR3, TLR7, TLR8 and TLR9 are involved in viral infection recognition. TLR5 and, less markedly, TLR2 and TLR4 play the leading role in the formation of symbiotic microbiota structure in the large intestine (Pang and Iwasaki 2012). As a result of various microbial ligand interaction with these receptors, signals are induced that are crucial for immune response to inflammatory effectors and for the activation of cell and innate immunity humoral components (Laino et al. 2016; Lebeer et al. 2018; Pang and Iwasaki 2012).

Low molecular peptides released from microbial cells in course of their life-sustaining activity can stimulate the development and functions of T helper (inducer) phenotype lymphatic cells (Buffie and Pamer 2013; Shaikh and Sreeja 2017). Reacting with Toll-like receptor 5, bacterial flagellins (protein component of microbial flagella) take part in the activation of gut mucosal immunity, in the chemotaxis of both pathogenic and symbiotic bacteria. DC localized in the gut mucus layer respond to bacterial flagellin effect with adhesion to host tissue cells and rapid production of chemokines, antimicrobial peptides and cytokines involved in immune response initiation. In response to flagellin effect, CD103+ DC produce IL-23 and partly induce IL-22 cytokine synthesis by innate lymphoid cells (ILCs), which stimulates protective immune mechanisms of the host related to epithelial cells (Kinnebrew et al. 2012; Shaikh and Sreeja 2017; Singh et al. 2018).

The interaction of bacterial DNA containing non-methylated cytosine phosphate guanosine dinucleotides (CpG) with Toll-R9 allows symbiont microbes to act as a local immune response adjuvant. Peptidoglycans of symbiotic (probiotic) bacteria that pass in considerable amounts from intestinal lumen into blood contribute to enhanced immune response of bone marrow neutrophils with respect to pneumococci and staphylococci. The presence of polysaccharide *A (PSA) Bacteroides fragilis* in a free state stimulated Tregs differentiation, IL-10 production and protected mice from experimental colitis caused by *Helicobacter hepaticus*; it was connected with *PSA* capability to increase the number of Tregs producing this interleukin. Skeleton *P-CWS* (the structure of cell wall skeleton consisting of peptidoglycane, isopentadecanoic acid and acidic polysaccharide) isolated from *Propionibacterium acnes* had anti-tumor activity due to the activation of macrophage cytotoxic functions. LPS of gram-negative symbiotic bacteria *in vitro* stimulated DNA synthesis in B cells in absence of T cells and macrophages. Probiotic strains of lactobacilli inactivated by heating triggered different immune responses, which is attributed to microbial pro- and anti-inflammatory mediator ratio (surface and secreted proteins and metabolites) capable to induce cytokine secretion as a result of NF-κB activation in monocytic cells. Cell-free filtrate of these lactobacilli cultural fluid inhibited the production of pro-inflammatory cytokines, signaling immune cells, tumor necrosis factor (TNF). One of the possible lactobacilli metabolites capable of suppressing TNF production in human myeloid cells is histamine, which is formed in course of microbial transformation of L-histidine amino acid (Engevik and Versalovic 2017; Ganesh and Versalovic 2015; Lebeer et al. 2018; Shaikh and Sreeja 2017). Surface components of probiotic lactobacilli cell walls induced cytokine production by type 1 and type 2 helper T cells in healthy donor blood monocytes. TGF-β secretion increased most markedly, while IL-10 and TNF-α secretion increased to lesser extent. In the same model, cell-free filtrates of cultural fluid of 17 lactic acid bacteria strains including probiotic cultures helped increase the production of IL-10 and decreased the formation of IL-2, TNF-α and IL-4; under such conditions the production of IL-12, IFN-γ and TGF-β stopped completely. Based on these data, the authors (Aguilar-Toaláa et al. 2018; Ashraf et al. 2014; Engevik and Versalovic 2017; Legent and Norris 2009) came to the conclusion that anti-inflammatory TNF-α immune responses are primarily related to the effect of probiotic lactobacilli surface structures on peripheral blood mononuclear cells, while anti-inflammatory immune effects are caused by joint action of microbial metabolites and bacterial surface components. Moreover, symbiotic (probiotic) bacteria produce various bacteriocins, microcins and other antimicrobial molecules that can indirectly modulate the immune functions by suppressing the growth and virulent potential of pathogenic and opportunistic microorganisms having intruded into the macroorganism (Engevik and Versalovic 2017; Lebeer et al. 2018; Mimee et al. 2016; Shaikh and Sreeja 2017; Singh et al. 2018).

Various probiotic microorganisms, their structural components (peptidoglycans, exopolysaccharides, LPS, nucleic acids, S-layer proteins etc.) and metabolites (various peptides, SCFA, homoserine lactones, dopamine, serotonin, histamine etc.) have specific receptors and targets in different immune system components and cells. Immunologic effects of probiotics and their low molecular effectors can differ in mammals and humans, in a healthy organism and in immune-compromised people with inflammatory manifestations (Engevik and Versalovic 2017; Laino et al. 2016; Nakagawa and Miyazaki 2016; Oleskin and Shenderov 2016). Low molecular weight compounds and cell components linked to cell walls, membranes and other microbial organelles of probiotic bacteria released during bacteriolysis, metabolites and signaling molecules (exopolysaccharides, membrane fragments, lectins, glutathione peroxidase, superoxide dismutase, nicotinamide adenine dinucleotide oxidase and peroxidase, organic acids, uronic acid etc.) can exhibit antioxidant and signaling activity (Shenderov and Lakhtin 2004; Aguilar-Toaláa et al. 2018; Engevik and Versalovic 2017; Lakhtin et al. 2007; Shaikh and Sreeja 2017; Singh et al. 2018; Wang et al. 2017). Also, the representatives of symbiotic gut microbiota form a multitude of various low molecular neuromediators capable of modifying the total pool of these regulators of nervous system functions in the human organism (Table 1).

Along with various eukaryotic human cells, the digestive tract microbiota takes an active part in the formation of the total pool of gas neurotransmitters (nitrogen oxide, carbon monoxide, hydrogen sulfide, ammonia, hydrogen) (Shenderov 2015; Nicolson et al. 2016; Oleskin and Shenderov 2016), oxytocin, vasopressin, neuropeptides, brain-derived neurotrophic factor and other hormones and neuroactive compounds in mammal organisms (Sampson and Mazmanitan 2015).

Table 1 Some neuromediators formed by the representatives of human symbiotic microbiota (Oleskin et al. 2016; Carabotti et al. 2015; Clarke et al. 2014; Cryan and Dinan 2015; Engevik and Versalovic 2017; Lebeer et al. 2018; Maguire and Maguire 2019; Oleskin et al. 2014; Oleskin and Shenderov 2016; Sampson and Mazmanitan 2015; Singh et al. 2018)

Neuromediators	Microorganisms
Serotonin	lactococci, lactobacilli, streptococci, *E. coli*, *Morganella*, *Hafnia*
C-dihydroxyphenylalanine (DOPA)	lactobacilli
Dopamine / noradrenaline	bacilli, lactobacilli, *E. coli*, staphylococci, *Hafnia*
Noradrenaline	bacilli, *E. coli*, *Serratia*, proteidae
Acetylcholine	lactobacilli
GABA	lactobacilli, bifidobacteria
Histamine	lactococci, lactobacilli, streptococci, *Morganella*, *Klebsiella*, *Hafnia*
Tryptamine	clostridia, ruminicocci
Tyramine	lactobacilli, enterococci
Aspartate, glutamate, glycine, taurine, tryptophan	lactobacilli
Butyrate, acetate, propionate, GABA	lactobacilli, bacteroides, clostridia, bifidobacteria, *Roseburia*, eubacteria

Over the course of many decades, the dominating opinion was that health, longevity and disease risk are related to mutative changes in DNA linear structure. In the modern view, the genetic system should be supplemented with the epigenetic one, which regulates activation and silencing of microbial and eukaryotic cell genes in response to exogenous or endogenous influence of different effectors. Epigenetic changes arise as a result of covalent attachment/detachment of different chemical groups (methyl, acetyl, phosphate etc.) to DNA, histones and other proteins. Epigenetic alterations are observed 100 times as often as traditional ones. Epigenetic changes in DNA and chromatin are reversible, but they can persist over several cell generations. Throughout the life of an individual the spectrum of “active and silent genes” may change.

There are many factors and agents (Fig. 1) that can interfere with epigenome regulation of gene expression and post-translational modification of gene products (Kanherkar et al. 2014). Among different modulators of epigenetic mechanisms responsible for the expression of chromosomal, mitochondrial and microbial genes throughout the life of an individual, symbiotic microorganisms have an important role to play (Shenderov 2013c; Paul et al. 2015; Shenderov 2016b; Singh et al. 2018). It is known that epigenetic biochemical machinery is closely connected with energy processes in the organism. According to the latest data, mitochondria and membranes of symbiotic microorganisms should be considered as a single metabolically active collective “organ” responsible for energy synthesis in the host organism and production of compounds regulating the expression of nuclear, mitochondrial and microbial genes. Energetic and other processes in mitochondria and their functional analogs (inner bacterial membranes) require a multitudes of

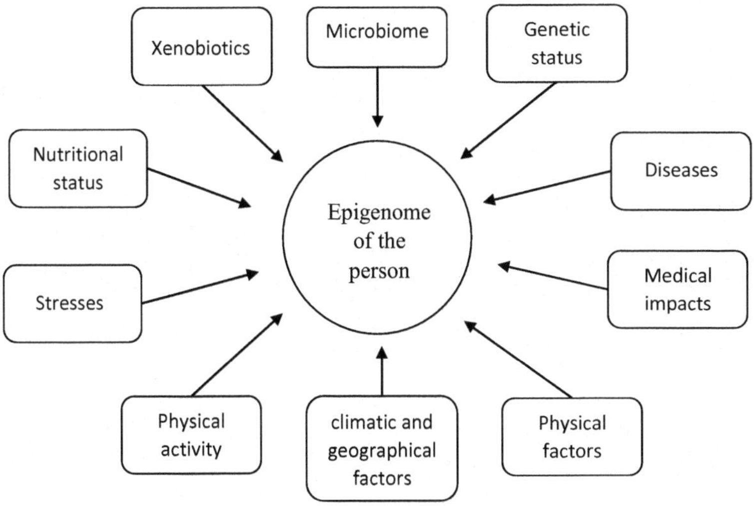

Fig. 1 Exogenous and endogenous factors involved in the formation of human epigenome

enzymes and their co-factors: vitamins (B1, B2, B3, B5, B6, B7, B9, B12, C, K1, K2), non-vitamin substrates (NAD, NADH, NADP+, NADPH, ATF, cytidine triphosphate, S-adenosyl methionine, 3'phosphoadenosine-5'phosphosulfate, glutathione, coenzyme B, coenzyme M, coenzyme Q10, co-factor F-430, heme, alpha-lipoic acid, methanofuran, molybdopterin/molybdenum co-factor, pyrroloquinoline quinone, tetrahydrobiopterin, tetrahydromethanopterin), minerals (*Ca, Cu*, Fe^{++}, Fe^{+++}, *Mg, Mn, Mo, Ni, Se, Zn*), amino acids (arginine, lysine, methionine, cysteine, β-alanine, serine, threonine, histidine, tryptophan, aspartic acid, carnitine), organic acids, Krebs cycle participants, certain nucleotides (for instance, pyrimidine), microRNA etc., which can be of microbial, dietary or endogenous origin (Shenderov 2016b; Shenderov and Midtvedt 2014).

Food products and symbiotic microorganisms have their effects on energy metabolism, human behavioral, metabolic and signaling reactions by supplying the organism with dietary substrates, co-factors, as well as through neuroendocrinal, metabolic, epigenetic and immune mechanisms. Long-term and profound functional impairment of the bacterial axis – the mitochondrion – epigenetics is accompanied with insufficient production of adenosine triphosphate and delayed delivery of this energy source to cells and tissues, sharp increase in the cells of high-reactive free radicals of oxygen, nitrogen and peroxidation products, antioxidant defense deficiency, disturbance of apoptosis and other cell development cycles, changes in signal transduction, proteostasis, cellular senescence, telomere length shortening, diminishing of stem cell pool, modification of balance of the compounds involved in the regulation of gene expression processes in mitochondrial, bacterial and nuclear DNA, in post-translational modification of the end products of these genes. It follows from the above that low molecular components of food and functional ingredients of symbiotic microbiota can be viewed as key players in the operation of energetic and epigenetic biochemical machinery (Shenderov 2008; Shenderov 2015; Shenderov 2016a; Bhat and Kapila 2017; El Aidy et al. 2016; Galland 2014; Sampson and Mazmanitan 2015; Shenderov and Midtvedt 2014; Singh et al. 2018; Stilling et al. 2014). Epigenetic changes that arise in mammal organisms throughout the life of animals and humans can be passed on to successive generations resulting in a growing disposition toward stress response defects and the development of a multitude of chronic metabolic diseases (Shenderov 2016a; Shenderov 2016b).

When discussing the importance of symbiotic (probiotic) microorganisms in the formation of human metabolome, it should be mentioned that the microbiome contains a cluster of genes responsible for the synthesis of a number of specialized microbial metabolites (polyketides, sterols, isoprenoids, nonribosomal peptides, other natural products) whose functions are currently underexplored; it is beyond dispute that such microbial molecules can potentially take part in a number of important immune, metabolic and regulatory reactions of the macroorganism. It will suffice to mention that such a special microbial metabolite as rapamycin is currently being clinically tested as means of treatment and prophylaxis in case of diseases connected with impairment of immune system functions. Undoubtedly, in perspective, the number of similar microbial metabolites with specific targets will only be growing (Dorrestein et al. 2014). In structure and function, microbial compounds often represent analogues or homologues of dietary substances and/or produced endogenously by host eukaryotic cells. It is safe to say that low molecular weight compounds serve as global (universal) regulators of intra- and interpopulation informational communication of any living organisms irrespective of evolutionary organization level (Lebeer et al. 2008; Lebeer et al. 2018; Shenderov 2011a; Shenderov 2013a).

Drawbacks and Negative Consequences of Traditional Probiotics Based on Live Microorganisms

The growing market of pharmaceuticals, dietary supplements, and particularly food products based on live probiotic microorganisms, is evidence of their sufficient prophylactic and, if less marked, therapeutic effects. However, it is currently impossible to precisely determine the optimum quantity of bacteria required for beneficial probiotic effects on human health; in many cases exact data are lacking on probiotic effect mechanisms and targets. It turned out that positive effects of live probiotic microorganisms may be short-term, uncertain or even absent in case of long use. The absence of clear health-promoting effect of prescribed probiotics on the base of live organisms is usually explained by low concentrations of biologically active microbial compounds in places of their application (Reid et al. 2011). In the past few years, experimental and clinical studies have also shown the difficulty, and sometimes even impossibility, of constructing industrially manufactured probiotics for targeted maintaining optimum levels of indigenous microbiota by way of simple selection of a particular live probiotic strain or even their group (Shenderov 2011a).

In the last 10–15 years, our knowledge regarding the safety of probiotics on the base of live microorganisms has changed considerably. And though the use of such probiotics for over 50 years has shown that they are quite safe as a means of disease prevention and treatment and as food products, it turned out that some probiotic bacterial strains can be harmful even if they belong to *Lactobacillus* or *Bifidobacterium* genera whose representatives have no traditionally known genes of pathogenicity. The accumulated documentary evidence shows that lactobacilli and bifidobacteria used in the production of fermented milk products or probiotics can act (if rarely) as agents of opportunistic infections (endocarditis, sepsis, bacteraemia, pneumonia, abdominal abscesses, meningitis, urological infections), especially in patients receiving antibiotic treatment or those severely immune compromised (Bourdichon et al. 2012; Cannon et al. 2005; Ohishi et al. 2010; Van Reenen and Dicks 2011; Yazdankhah et al. 2009). The same bacteria may be responsible for allergic or autoimmune pathologies (Berer et al. 2011; Guilherme

B. A. Shenderov et al., *METABIOTICS*,
https://doi.org/10.1007/978-3-030-34167-1_9

et al. 2006; Kiseleva et al. 2011; Prangli et al. 2010). Symbiotic microorganisms including probiotic strains may increase platelet aggregation aggravating clinical manifestations of hemolytic uremic syndrome (Van Reenen and Dicks 2011; Yazdankhah et al. 2009); some of them may be a source of toxic biogenic amines (Bourdichon et al. 2012). For example, in the process of microbial transformation of phosphatidylcholine contained in red meat and cheese, choline reacts to form trimethylamine which oxidizes to trimethylamine -N-oxide — a compound that takes part in atherosclerotic plaque formation; it also plays a role in thrombosis development, certain hepatic lesions and colon cancer (Gilbert et al. 2016; Sonnenburg and Backhed 2016).

The majority of strains used as starter cultures in the production of probiotics have been selected according to such characteristic as high antagonistic activity (against bacteria or fungi). Once in the gastrointestinal tract or on female genital mucosa, these probiotic microorganisms can suppress the growth and development of symbiotic gut and vaginal microbiota and modify its metabolic activity (Glushanova and Shenderov 2005; Shenderov and Glushanova 2006; Yazdankhah et al. 2009). Furthermore, some gut bacteria can transform orally administered pharmaceuticals, modifying their activity and/or transforming inactive pro-drug forms into a pharmacologically active drug (Clayton et al. 2009). Unfortunately, we have not found any data in literature on *in vitro* and *in vivo* research into the capability of probiotic microorganisms to interfere with drug bioaccessibility and pharmacological activity; but such effects of probiotics cannot be excluded. The situation becomes even more uncertain if probiotic strains belong to *Enterococcus, Streptococcus, Escherichia, Bacillus, Coprococcus, Bacteroides* or other genera containing several strains or created on the basis of genetically modified microorganisms (Shenderov 2013a).

Data is available that certain probiotics (*Lactobacillus rhamnosus* GG strain) can induce additional expression of 400 new genes in the gut which are involved in immune response and inflammation, cell growth and differentiation, apoptosis, intracellular information exchange and adhesion processes (Di Caro et al. 2005). Oral administration of a probiotic containing live *E. faecalis* bacteria changed the activity of 42 human genes involved in cell cycle regulation, apoptosis and intermicrobial exchange (Cancer-causing gut bacteria exposed 2018). Moreover, it was found that when passing the digestive tract, some silent genes of probiotic microorganisms can also get activated; unfortunately, many newly formed bacterial products have never been extensively characterized from both the chemical and the functional perspective (Bron et al. 2004; Di Caro et al. 2005; Yuan et al. 2008). So, these were the conditions for the induction of 72 genes of a probiotic strain *L. plantarum* WSFS1 to occur. Nine genes among them are responsible for the transport and enzymatic fermentation of sugars, another nine — for the synthesis of amino acids, nucleotides, vitamins and other co-factors, four — for the synthesis of surface proteins controlling the synthesis to the specific immune factors of the host; one more process that took place in this probiotic *Lactobacillus* strain was the induction of expression of 46 more genes; chemical and protein products of these genes had

Table 1 Some ecological and social consequences of the use of probiotics on the base of live microorganisms and fecal content transplantation (Tkachenko 2014; Chernevskaya and Beloborodova 2018; Shen et al. 2018; Shenderov 2007)

Item No.	Some consequences of probiotic use
1	Spreading of drug-resistant microorganisms
2	Decreased effectiveness of chemotherapy and chemoprophylaxis; increased cost of treatment for many diseases
3	Selection of microorganism strains with atypical biological characteristics
4	Formation of new unconventional microbial associations
5	Changes in pharmacokinetics and biotransformation medicines and food nutrients
6	Changes in etiological structure of infectious diseases
7	Broadened spectrum of diseases related to microbial factor
8	Increased number of people with low resistance to infectious agents
9	Increased number of people with altered mental and behavioral responses

not been identified (Bron et al. 2004). Thus, the expression of known or silent genes in microbial and/or eukaryotic cells when consuming live probiotic microorganisms may have an undesired or even unpredictable effect on human health. In the last 10 years, evidence was gathered (Table 1) showing that lactic acid and other probiotic bacteria are involved in horizontal gene transfer of antibiotic resistance genes, traditional and new pathogenicity factors and genes controlling other functions (hemolysis, D-galactose, DNA-methyltransferases, sirtuines, glucuronidase, acetaldehyde, pks genes etc.) by way of transformational, transduction and conjugal DNA transfer. Gene transfer and recombination processes related to the administration of live probiotic microorganisms can involve far-reaching potentially negative health and environmental effects (Bourdichon et al. 2012; Sommer et al. 2010; Van Reenen and Dicks 2011).

Various experimental animal models, as well as clinical observations have shown that live probiotic microorganisms can modify the functions of the NF-κB signaling system which regulates the activity of genes of synthesis of anti-inflammatory interleukins, hemokinins and cytokines (especially, IL-6, IL-8, adhesive and other molecules). They also lead to impaired regeneration of liver and gut cell lesions, cause telomere disfunction which results in increased risk of premature ageing. Moreover, it was found that DNA-methyltransferases of *E. coli* can cause potential damage to epigenetic integrity of cell genomes. In the process of chromosomal DNA sections methylation, DNA-specific microbial DNAses can cause cell death resulting from microbial cleavage of eukaryotic cell DNA at sites where methyl groups are added (Ishikawa et al. 2011). There are indications that some *E. coli* strains (including probiotic strain *E. coli Nissle* 1917) possess a set of genes (*pks* island) that are responsible for the production of double-strand breaks in eukaryotic host cell DNA. Bacteria containing the pks genes induce in eukaryotic cells a process (referred by the term "megalocytosis") in which the cell body and nucleus become enlarged and mitosis stops. A break in a double-strand DNA of eukaryotic cells occurs after their four-hour contact with bacteria carrying *pks* islands. Over 30% of *E. coli* strains isolated from healthy adult feces had such *pks* islands.

The speed of DNA damage of eukaryotic cells in the gut depends on the number of *E. coli* contacting host cells. The exposure of the latter to 100 or more bacteria, a great number of lesions develop in DNA structure of epithelial intestinal cells. Such DNA breaks caused by microbial peptide-polyketide genotoxins formed by certain symbiotic (probiotic) bacteria can trigger various disorders in an animal organism including development of neoplasms (Arthur et al. 2012; Nougayrede et al. 2006). Live cells of *E. faecalis* inside the intestinal tract of mammals release substantial amounts of extracellular superoxide, hydroxyl radicals and hydrogen peroxide during carbohydrate fermentation and via autoxidation of membrane dimethylmenaquinone. These oxidants can damage DNA of enterocytes and facilitate the development of sporadic adenomatous polyps and colorectal cancer (Huyckel et al. 2003; Cancer-causing gut bacteria exposed 2018). Ethanol and its metabolite acetaldehyde are classified as class I carcinogens. Acetaldehyde concentrations between 100 and 500 μM can cause mutagenic damage in the structure of eukaryotic cell DNA. Many symbiotic microorganisms including lactic acid bacteria used as starter cultures in the production of fermented food products and in probiotics can convert ethanol and/or glucose to carcinogenic acetaldehyde. The levels of this compound contained in foods or in the gut can exceed the values required for the manifestation of acetaldehyde's mutagenic effect. Moreover, as live probiotic microorganisms can colonize the digestive tract for quite a long time and produce considerable amounts of acetaldehyde locally, they can be potentially more harmful for health than traditional dairy lactic acid bacteria (Helminen et al. 2011; Nummi et al. 2011). Thus, probiotics on the base of various microorganisms have a wide range of potential drawbacks. This calls for more thorough research into the efficiency and safety of probiotics – both already produced commercially and being developed. It is especially important when it comes to consuming such probiotics by pregnant women, neonates, infants, as well as people with various immunodeficiencies. Recently published data have presented convincing evidence that transitory colonization of pregnant germfree mice led to considerable changes in the functions of small intestine immune cells in newborn mice (Charbonneau et al. 2016). It follows that the attempts to use probiotics produced by traditional technologies for manipulating microecological, immune, metabolic and epigenetic status in pregnant and breastfeeding women may have a considerable influence on neonatal development of their offsprings (Charbonneau et al. 2016; Jurk et al. 2014; Shenderov 2013a; Yazdankhah et al. 2009).

The analysis of the published data (Green et al. 2017; Sanders et al. 2016) allows to arrive at a conclusion that currently available knowledge related to the effectiveness and safety of traditional probiotics made with live microorganisms are not enough for reliable assessment of their ecological and clinical risk in short- and long-term perspective. The above mentioned shortcomings of live probiotic microorganisms had led *European Food Safety Authority (EFSA)* to conclude on 4

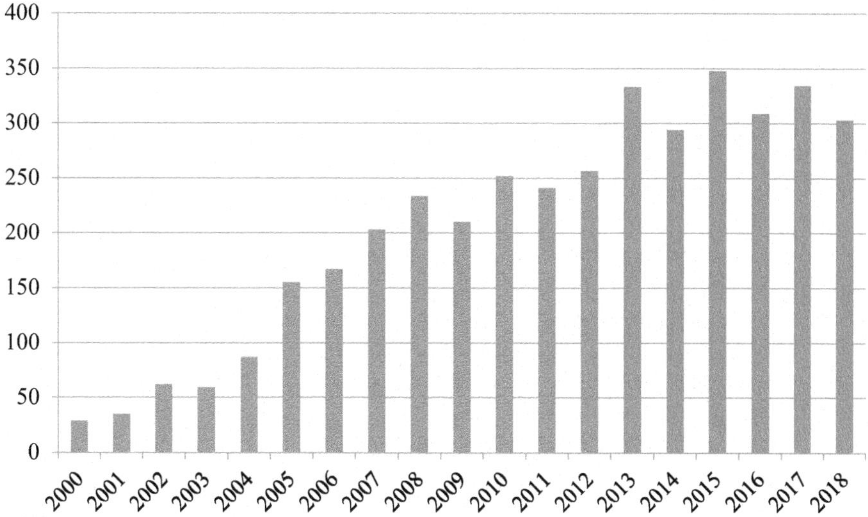

Fig. 1 The number of publications on clinical trials of probiotics according to PubMed.gov

December 2012 that a ban is necessary for using designations (labels) on products containing live probiotic microorganisms which state that using these foods has positive health effects. From the perspective of *EFSA* experts, there is not sufficient clinical proof for their health benefits (Katan 2012). A similar ban was imposed by *Food & Drug Administration (FDA)* of the USA in 2014 (Frequently Asked Questions About Medical Foods 2016). In course of the workshop *"Science and Regulation of Live Microbiome based Products: no headway on regulatory issues"* held by the *FDA Center for Biologics Evalution and Research and Nutritional Institute of Allergy and Infectious Diseases"* (Rockville, USA, September, 2018) it was stated that many issues related to safety of probiotic products call for extra decisions. In some US hospitals, it is forbidden to buy, store, prescribe and distribute probiotics to prevent contamination of wards and patients with live probiotic microorganisms. In recent years, there has been a tendency of reduce the number of publications devoted to clinical studies of probiotics (Fig. 1).

As an alternative to traditional probiotics, prebiotics, dietary fibers and other compounds had been offered which selectively stimulate the growth and/or activity of one or a limited number of representatives of genus (genera)/species in gut microbiota having positive effects on the host organism (Gibson and Roberfroid 1995). By now, the prebiotic range has grown dramatically to include several dozens of alcohols, oligosaccharides, polysaccharides, enzymes, peptides, plant and microbial extracts and other compounds (Table 2). Their production volumes have reached hundreds of thousands tons annually.

Table 2 Basic types of prebiotic substances (Doronin and Shenderov 2002; Shenderov 2008; Shenderov 2014b; Maguire and Maguire 2019; Roberfroid et al. 2010)

Item No.	Substances
1	Monosaccharides, alcohols (xylitol, melibiose, sorbitol, raffinose etc.)
2	Oligosaccharides (lactulose, galacto-oligosaccharides, fructo-oligosaccharides, soybean oligosaccharides — stachyose etc.)
3	Polysaccharides (pectin, inulin, chitosan, pullulan etc.)
4	Enzymes (microbial galactosidases, proteases in Saccharomycetes etc.)
5	Peptides (of soybean, milk etc.)
6	Amino acids (valine, arginine, cysteine, glutamic acid etc.)
7	Antioxidants (vitamins A, C, E, carotenes, glutathione, ubiquinone, selenium salts etc.)
8	Unsaturated fatty acids (omega-3 etc.)
9	Organic acids (butyric, propionic, acetic etc.)
10	Plant and microbial extracts (carrot, potato, tomato, rice, garlic, yeast etc.)
11	Other (lecithin, para-aminobenzoic acid, lysozyme, lactoferrin, gluconic acid, starch syrup etc.)
12	Prebiotics on the base of microbial-origin polysaccharides

Metabiotics: New Stage of the Probiotic Concept Development

The knowledge of molecular language of symbiotic (probiotic) microorganisms allows for a more intensive and targeted development of the next generation of probiotics and functional foods. In our opinion, the development of a traditional probiotic concept includes the creation of therapeutics, dietary supplements and functional foods that can be denoted by the common term of "metabiotics" (Shenderov 2007; Shenderov 2009; Shenderov 2011a; Shenderov 2011b). We suggest to understand by the term all microecological products designed to preserve and restore the content and functions of symbiotic microbiota in humans, animals and plants and based on water soluble components of cells, metabolites and signaling molecules secreted or released at the time of microbial cell destruction in known or potential probiotic strains of microorganisms, as well as synthetic (and/or semisynthetic) metabiotics which will be artificially constructed as analogues or improved copies of natural biologically active compounds formed by symbiotic microorganisms. Metabiotics have a known chemical structure and their application allows to optimize organism-specific epigenetic, physiological functions, regulatory, metabolic and/or behavioral reactions related to the activity of host indigenous microbiota. This new group of microecological products should be equal to or surpass the first-generation probiotics in their therapeutic and prophylactic efficiency, while being safer than traditional probiotics.

It is worth noting that the idea of using symbiotic (probiotic) microorganisms as starter cultures for the production of low molecular weight bioactive compounds was discussed in scientific literature by many researchers. For such microbial products, apart from the term "metabiotics" suggested by us over 20 years ago (Shenderov et al. 1997; Shenderov 2007), such terms were used as "killed by heating probiotics" (Indriyani et al. 2012), "metabolite probiotics" (Vakhitov and Sitkin 2014; Vakhitov et al. 2005), "postbiotics" (Neish 2009; Tsilingiri and Rescigno 2012), biological drugs (Sonnenburg and Fischbach 2011), "pharmabiotics" (Caselli et al. 2011; Sobol 2017), "paraprobiotics" (Almada et al. 2016). Over the past few years, functional products and pharmaceuticals for metabolite therapy of diseases related to human microbiota imbalance have been most frequently denoted in clinical and

B. A. Shenderov et al., *METABIOTICS*,
https://doi.org/10.1007/978-3-030-34167-1_10

scientific literature by the term "metabiotics" (Ardatskaya 2015; Ploskireva 2016; Shenderov 2017; Shaikh and Sreeja 2017; Sharma and Shukla 2016; Shenderov 2011a; Shenderov 2011b; Singh et al. 2018), "postbiotics" (Aguilar-Toaláa et al. 2018; Maguire and Maguire 2019; Sobol 2017) and "pharmabiotics" (Sobol 2017). Known and new potential strains of symbiotic (probiotic) microorganisms may become a source of hundreds or even thousands of low molecular weight bioactive compounds (bacteriocins and other antimicrobial molecules, short-chain and other fatty acids, biosurfactants, polysaccharides, peptidoglycans, teichoic acids, lipo- and glycoproteins, vitamins, antioxidants, nucleic acids; proteins including enzymes and lectins; peptides with different functional activity, amino acids, growth and coagulation factors, inducers of defensin-like molecules, various signaling and transport molecules, plasmalogens, phenol-containing compounds, co-factors etc.) (Shenderov et al. 2010; Aguilar-Toaláa et al. 2018; Almada et al. 2016; Caselli et al. 2011; Engevik and Versalovic 2017; Holmes et al. 2011; Lebeer et al. 2008, 2010; Shenderov 2013a; Singh et al. 2018; Sonnenburg and Fischbach 2011). Various strains of symbiotic (probiotic) microorganisms can form sets of low molecular weight compounds, and are thus candidates for the construction of general-purpose or specific metabiotics. Real effectiveness and safety of these microbial molecules and metabiotics developed on their basis will largely depend on the physical and chemical characteristics of microbial bioactive components, their interrelationship and relations with other components at the sphere of application, which can both intensify and inhibit their activity, competition in the struggle for specific transport proteins or absorption site. The physiological state of the consumer or his/her disease, diet, prescribed medicine can considerably modify the effectiveness and safety of microbial biologically active substances (metabiotics). Some groups of microbial low molecular weight compounds formed by probiotic microorganisms (SCFA, other organic acids, various proteins, peptides, amino acids, nucleic acids, nucleotides, microbial polysaccharides, peptidoglycans, vitamins, antioxidants; various co-factors — coenzyme A, coenzyme Q; various transport and signaling molecules including gas and other neurotransmitters) are currently being used in industrial production of metabiotics (Shenderov 2011a; Shenderov 2013a; Singh et al. 2018).

For some metabolites produced by symbiotic (probiotic) bacteria, targets and effects have been set in prokaryotic and eukaryotic cells. Low molecular weight microbial compounds formed from endogenous and exogenous substrates can either pass freely through the wall and membranes of various eukaryotic and microbial cells, or else they need specific receptors and transporters for penetration. They can act as substrates, co-substrates, metabolites in different biochemical and signaling reactions; some of them possess either metabolic or only signaling activity. Depending on their application site, these bioactive molecules are divided into those with molecular-level effects (gene replication and expression, transcription and translation of genetic information; post-translational effects), cell-level effects (on the surface, in cell membranes; biosynthesis of protein and energy in mitochondria and ribosomes), inside cell hyaloplasm (in location sites of the nucleus, organelles and cell inclusions), in extracellular matrix (at location sites of capillaries, synapses of nerve endings, performance of metabolic reactions, extracellular information exchange etc.), in tissues, organs, physiological systems and the organism on the whole (Aguilar-Toaláa et al. 2018; Engevik and Versalovic 2017; Shenderov 2011a; Singh et al. 2018). The targets for microbial bioactive compounds in mammal organisms may be represented by the metagenome and meta-epigenome of eukaryotic and microbial cells, epigenetic control of particular gene expression and post-translational modification of gene products, intra- and intercellular information exchange among microbial populations between the host and its symbiotic microbiota, including quorum-sensing regulation, metabolic and behavioral reactions, growth and development of the superorganism and its health on the whole.

Methods and Techniques Used for Obtaining and Identifying of Microbial Low Molecular Weight Cellular Compounds, Metabolites and Signaling Molecules

Different techniques and analytical technologies are used for identification of compounds potentially suitable for constructing various metabiotics. Their choice depends on analytical purposes, qualitative and/or quantitative characteristics of microbial complexes and molecules under study. Table 1 shows some techniques used for microbial cell destruction, removal of live microorganisms and extraction of low molecular weight biologically active compounds'.

Modern techniques and methods of metabolomics allow to identify low molecular weight structural cellular compounds, metabolites and signaling molecules in source raw materials (cellular microbial suspensions and low molecular weight compounds in cultural fluid purified from viable microbial cells), in ready-made products (metabiotics of various form and nature), in biological fluids (saliva, blood, urine, cerebrospinal fluid), as well as in human tissues and intestinal content. Human gut microbiota possesses metabolic activity comparable with that of the liver (Beger et al. 2016; Yadav et al. 2018). In the last 15–20 years, numerous analytical techniques and technologies had been developed for metabolome profile study and evaluation in different samples, which found their way into practice (Table 2). In countries with advanced industrial economies (the USA, UK etc.) phenotype centers have been established and are running, capable of annually analyzing many hundreds of thousands of various biological material samples and simultaneously identifying dozens, hundreds and thousands of low molecular weight molecules.

Various kits have been developed and find an increasingly greater application for the analysis of antibiotics, specific nutrients, enzymes, co-factors and other molecules involved in any given metabolite pathway of prokaryotic and eukaryotic cells. The *Biocrates Life Sciences* company sells test systems for measurement and identification of a limited range of metabolites, amino acids, biogenic amines, 150 lipids and 16 bile acids (Beger et al. 2016; Biocrates Life Sciences. http://www.biocrates.com/products/research-products). The combination of a great number of technologies and platforms allows to evaluate a wide range of metabolite profiles in different biological materials.

B. A. Shenderov et al., *METABIOTICS*,
https://doi.org/10.1007/978-3-030-34167-1_11

Table 1 The techniques of removal of live starter microorganisms, destruction of microbial suspensions, isolation of complex low molecular weight biologically active compounds, microbial metabolites and signaling molecules (Aguilar-Toaláa et al. 2018; Almada et al. 2016; Birmpa et al. 2013; Diels and Michiels 2016; Efaq et al. 2015)

Item No.	Techniques
1	Mechanical destruction
2	Frequent freezing and thawing
3	Use of enzymes with specific activity against microbial cell walls, membranes, intracellular structures
4	Microwave radiation
5	Supercritical water extraction
6	Ionizing radiation
7	Ultraviolet radiation
8	High hydrostatic pressure
9	Ultrasound damaging
10	Cell damaging by chemical agents (for example, exposure to acids)
11	Centrifugation at different rotor revolutions
12	Different kinds of dialysis, the use of ion exchange resins
13	Different techniques of ultrafiltration of whole and destroyed cell suspensions
14	Inactivation of microorganism suspensions and extraction of cell components and metabolite compounds using liquid carbon dioxide

Thermal effects on microbial cell suspensions (60°C, 30 min; 80°C, 20 min; 95°C, 5 min; 100°C, 5 min)

Table 2 Some metabolome technologies used for content evaluation and identification of low molecular weight compounds that define the metabolome of the corresponding biological sample (Aguilar-Toaláa et al. 2018; Aurich and Thiele 2016; Beger et al. 2016; Braune and Blaut 2016; Nicholson et al. 2012; Yadav et al. 2018)

Item No.	Technologies
1	Gas chromatography
2	Liquid chromatography
3	Mass-spectrometry
4	Nuclear magnetic resonance spectroscopy
5	High-efficiency liquid chromatography with subsequently obtained mass spectrum in negative and positive modes of operation
6	High-efficiency liquid chromatography with tandem high-resolution mass-selective detection with electrospray ionization
7	High-efficiency high-resolution liquid chromatography with tandem high-resolution mass-selective detection
8	Pulsed Field Gel Electrophoresis — PFGE
9	Capillary electrophoresis with mass-spectrometry-based detection
10	Two-dimensional gel electrophoresis

Low molecular weight bioactive compounds formed by commensal and symbiotic microorganisms of gut microbiota can easily penetrate the walls and membranes of eukaryotic and prokaryotic cells in the human organism; alternatively, they need preliminary interaction with specific receptors on the cell surface. According to effects on the corresponding targets, these bioactive microbial molecules may be divided into metabolic molecules, purely signaling molecules or molecules having at the same time specific structural, metabolic, signaling or another biological and pharmacological activity. As signaling molecules, they can activate, silence or modify the corresponding genes through the specific interaction with DNA receptors of human cells different in functionality by using kinases and/or other enzymes and other cell systems (Shenderov 2011a).

Low molecular weight bioactive compounds of symbiotic (probiotic) and commensal microorganisms display their effects on different levels of the organism:

1. Molecular level (gene replication and expression, transcription, translation of genetic information and its epigenomic modification).
2. Cellular level (surfaces, membranes, post-translational effects, energy and protein biosynthesis in mitochondria, ribosomes, exosomes).
3. Intracellular compartment — *gyaloplasme* (the location of nucleus, organelles, exosomes, various inclusions, ion channels).
4. Intercellular matrix (the location of capillary, neural synapses, metabolic reactions, synthesis and accumulation of hormonally active cell peptides, intercellular information exchange etc.).
5. Tissues, organs, physiological and metabolic systems.
6. Host organism as a whole.

The analysis of scientific literature data (Beloborodova 2012; Vakhitov et al. 2005; Sitkin et al. 2015; Aguilar-Toaláa et al. 2018; Dorrestein et al. 2014; Engevik and Versalovic 2017; Lebeer et al. 2018; Maguire and Maguire 2019; Singh et al. 2018), as well as our own experimental studies and clinical observations (Volkov et al. 2007; Vorobeichikov et al. 2008; Oleskin and Shenderov 2016; Shenderov 2013a; Shenderov 2015a; Shenderov and Lakhtin 2004; Shenderov et al. 2010; Shenderov et al. 2018a; Francavilla et al. 2008) have shown that, as a rule, *in vitro* cultural fluids of probiotic bacteria, as well as biological fluids (blood, urine, saliva etc.) of people receiving various probiotics orally, had very low levels of the quantitative content of target active microbial compounds. Unfortunately, traditional models and technologies used to determine the targets for application of such microbial bioactive compounds in nano- and micromolar concentrations and to objectively assess their beneficial health effects or negative side effects yield little information. New model systems sensitive to low concentrations of metabiotics should be found and tested.

Classification of Metabiotics and their Brief Description

Commercially available metabiotics can be divided according to their composition into dietary supplements, functional and personal foods, therapeutics, cosmetic products (liquid, dry, pastes, creams etc.); pharmaceuticals on the base of low molecular weight components related to structural components of microbial cells (Cellular metabiotics), low molecular weight metabolites and signaling microbial molecules (Metabolite metabiotics), metabiotics at the same time containing both microbial cell components and their metabolites and signaling molecules; hybrid multicomponent microbial components and metabolic and signaling molecules; metabolite formulations made with low molecular weight microbial components isolated from several microorganism strains or species; as well as hybrid multicomponent low molecular weight microbial and plant components and molecules. Also, semi-synthetic and synthetic metabiotics are currently being developed and launched on the food market. Metabiotics may be used both independently and as nutritional supplements to various known microecological therapeutics (Table 1) (Shenderov 2017; Shenderov et al. 2018a; Shenderov 2011a; Shenderov 2013a).

Depending on the nutritional base used for growing starter strains of symbiotic microorganisms, the obtained metabiotics may be differentiated into those of milk, plant and combined origin. Metabiotics may also be differentiated by method of extraction, inactivation of starter culture cells, delivery to the host organism, evaluation of stability and activity, the use of different technologies for their biological activity evaluation (in *in vitro* tests, on different kinds of experimental animals, in clinical trials).

B. A. Shenderov et al., *METABIOTICS*,
https://doi.org/10.1007/978-3-030-34167-1_12

Table 1 Potential forms of metabiotics

№	Forms of metabiotics
1	Metaprobiotics
2	Metaprebiotics
3	Metasynbiotics
4	Metanutritive supplements
5	Metabiotics

Some of the Best-known Metabiotics on the Market of Microecological Products

In the last decade, some metabiotics have been launched on the markets of a number of countries including the Russian Federation, whose production includes natural (or artificial) bioactive molecules similar or identical to those formed by the representatives of normal human microbiota. They have proven their therapeutic and prophylactic efficiency in some infectious and metabolic diseases (Table 1).

In addition, other metabiotics are available on the international market. Among them, of special note are pharmaceuticals which are essentially *E. coli* glycoprotein with anorectic activity (Tsuda et al. 1992); microbial complex of polysaccharide/glycopeptide of a *Lactobacillus casei* strain with hypotensive effect (Beider and Flambard 2003; Sawada et al. 1990); tripeptides (*Val-Pro-Pro* and *Ile-Pro-Pro*) of *Lactobacillus helveticus* strains which inhibit angiotensin-converting enzyme and mildly lowering blood pressure (Nakamura 2004; Nakamura et al. 2005). In the years to come, the metabiotics market will see the appearance of therapeutics and functional foods that contain proteins, peptides, adhesins (Lebeer et al. 2008; Lebeer et al. 2010), biosurfactants (Lakhtin et al. 2010), lectins (Shenderov and Lakhtin 2004; Lakhtin et al. 2007), nucleic acids and other cell wall components (Caselli et al. 2011) of lactobacilli, bifidobacteria, enterococci, as well as other symbiotic (probiotic) bacteria.

The manufacturers of metabiotics launched on the market as pharmaceuticals, dietary supplements or as functional food components are registered in accordance with national laws controlling the quality and safety of pharmaceuticals and food products. One of the key aspects in serial production of metabiotics on the base of symbiotic (probiotic) microorganisms of human origin for different applications, apart from isolation, taxonomic identification, research into positive target features and safety, is supplying biotechnological factories with standard high-quality starter cultures, or the so-called starter cultures for direct inoculation (SCDI), stable in their biological and technical properties. Traditional starters need preliminary incubation followed by a series of starter culture re-inoculations at the factory's premises for obtaining the required quantity of microbial cells prior to inoculating them

B. A. Shenderov et al., *METABIOTICS*,
https://doi.org/10.1007/978-3-030-34167-1_13

Table 1 Some of the best-known metabiotics (Belousova et al. 2005; Shenderov et al. 2018a; Aguilar-Toaláa et al. 2018; Shenderov 2013a; Singh et al. 2018)

Name	Contents and properties
Hylak forte (Germany)	Liquid pharmaceutical containing no microbial cells and intended for oral administration. It includes metabiotic products produced by the following strains: *Escherichia coli* DSM 4087 (25 g / 100 ml), *Streptococcus faecalis* DSM 4086 (12.5 g / 100 ml), *Lactobacillus acidophilus* DSM 4149 (12.5 g / 100 ml) and *Lactobacillus helveticus* DSM 4183 (50 g / 100 ml). The positive health effects of this metabiotic in children and adults is related to the restoration of gut microbiota content and functions, water-salt balance, acid-alkaline balance, normalization of vitamin B and K levels in microorganism, supplying enterocytes and immune intestinal cells with energy. The positive effects of this sterile liquid pharmaceutical are explained by the presence of volatile fatty acids, lactic acid and some unidentified low molecular microbial substances (Belousova et al. 2005; Kocharovec 2018; Rudkowski and Bromirska 1991)
Bactistatin (Russia)	Lyophilized cultural fluid of *Bacillus subtilis*, containing no live bacteria and including a complex of various low molecular weight compounds and metabolites and natural sorbent zeolite (*http://www.kkraft.ru*)
Colibiogen (Switzerland)	Filtrate of *Escherichia coli* (strain *Laves* 1931) cultural fluid, containing no proteins and including peptides, amino acids, polysaccharides and fatty acids (Zihler et al. 2009)
Pro-Symbioflor (Germany)	Microbe-free lysate of bacteria and cultural fluid of *Enterococcus faecalis*, *E.coli* (Klein et al. 2013)
Helinorm (Germany/ Russia)	Inactivated cells of probiotic bacteria *L.reuteri* that link specifically to *H. pylori* and remove these bacteria naturally from the organism; recommended for gastritis, gastric ulcer and cancer prevention (Hunt et al. 2011) (*http://www.kkraft.ru/en/#bs4*)
CytoFlora (USA)	Microbe-free cell wall lysates of *Bifidobacterium bifidum, B. infantis, B. longum, Streptococcus thermophiles, Lactobacillus reuteri, L. salivarius, L. casei, L. bulgaris* (Bioray. CytoFlora®. http://www.bioray.com/cytoflora/)
Del-Immune V (USA)	Peptidoglycans and DNA fragments of *Lactobacillus rhamnosus* (Del-Immune V. http://www.delimmune.com/research/)
Daigo (Japan)	A mixture of bioregulator peptides extracted from cultural fluids of 16 strains of various lactobacilli *L.curvatus, L.casei, L.acidophilus, L.plantarum, L.fermentum, L.salivarius, L.brevis, L.rhamnosus* after cultivating them on soymilk for a year (Popova et al. 2016)
Complex functional foods fortifier (Russia)	Cell-free filtrates of probiotic *Lactobacillus acidophilus NK-1* and *Bifidobacterium longum B 379 M* grown on hydrolyzed milk medium and purified from live bacteria by microfiltration; contains B vitamins, amino acids, macro- and microelements and organic acids. It is a complex fortifier, and the products resulting from its application are characterized by metabolic activity and positive effect on the human organism. Used as main or auxiliary component in production of various kinds of drinks, in the native or powder form or as a concentrate (Shenderov et al. 2010)
"New Class of Pharmabiotic" (Russia)	Obtained by long fermentation of raw materials (vegetables, herbs, fruit, berries, animal products) by lactobacilli (three strains) and *Streptococcus thermophilus* with marked proteolytic activity. After removal of unfermented sediment by centrifugation, peptides, amino acids, vitamins, microelements, proteases and volatile organic acids get accumulated in the liquid part. The resulting PP has multiple beneficial effects on human health (Sobol 2017; Sobol and Sobol 2012)

into industrial bulk containers. If quality standards are not properly applied, as a result of multiple re-inoculations, numerous microbial clones can appear in the original culture, which may be essentially different from the originally claimed starter strain as regards their technological and designated properties. One solution might be to introduce SCDIs i.e. bulk starters produced at specialized laboratories in the form of dry or frozen biomass of the producing strain; after delivery to the corresponding biotechnological factories, SCDIs are immediately inoculated into industrial bulk containers.

It will be promising to construct combined therapeutics (metabiotics + other known microecological substances), which will be launched in the market either as dietary supplements in their own right, or as new groups of functional foods fortified with these components (Shenderov 2013a; Sonnenburg and Backhed 2016).

Cellular Metabiotics and Metabolite Metabiotics

To provide a clearer picture of what modern techniques and technologies are used when developing metabiotics and launching them on the market of therapeutics and functional foods, let us describe how metabiotics were developed on the base of cell walls of the *Lactobacillus reuteri* DSM *17648* (Pylopass/Helinorm) and metabolites and signaling molecules on the base of the *Bacillus subtilis No.*3 probiotic strain (Bactistatin).

Cellular Metabiotics

Cellular metabiotics contains in its composition non-living probiotic cells. We will describe cellular metabiotics using the example of Pylopass/Helinorm.

Pylopass/Helinorm is a metabiotic product designed to control the *Helicobacter pylori* load in the stomach. *Helicobacter pylori* are gram-negative bacteria ever present in the stomach of humans and some mammals (for example, the dog, the cat). More than 50% of the population globally is infected with the bacterium *Helicobacter pylori*.

H. pylori is a major component in the causation of gastritis and *H. pylori* infection can also lead to gastric cancers and ulcers (Suerbaum and Michetti 2002). Today's strategy to manage *H. pylori* is based on antibiotics treatment with a regime called "triple therapy", which comprises two antibiotics and a proton pump inhibitor (Malfertheiner et al. 2017). This often leads to side effects and negative effects on our microbiota. The regime's efficiency is also becoming compromised by increasing rates of antibiotic resistance.

The probiotic strain *Lactobacillus reuteri* DSM17648 was developed as a metabiotic (a non-viable microbial preparation with health-beneficial properties) formulation (Pylopass) to manage stomach and gut health.

The chapter is written by Christine Lang (Technical University Berlin, Germany) and Kimmo Makinen (Novozymes S/A, Denmark).

B. A. Shenderov et al., *METABIOTICS*,
https://doi.org/10.1007/978-3-030-34167-1_14

Pylopass is a lactic acid bacterium of the species *Lactobacillus reuteri* which was specifically selected and developed to recognize *Helicobacter pylori* in the stomach and to co-aggregate with this bacterium. After binding, *Helicobacter's* motility is impaired and the co-aggregated *Lactobacillus–Helicobacter* complex is naturally removed by being flushed out of the stomach. This leads to a significant reduction of *H. pylori* load and offers a novel regimen to control *H. pylori* load in the stomach after triple eradication therapy and as a prophylactic supplement.

Helicobacter and Stomach Health

Contrary to the long-held belief that germs do not survive in the stomach, *Helicobacter pylori* (*H. pylori*) is found in the stomachs of more than 50% of the population worldwide. *H. pylori* is a Gram-negative microaerophilic, spiral bacterium that colonizes the gastric mucosa. It is able to produce the enzyme urease and thereby degrade urea to ammonia, which buffers against the acidic environment of the stomach. This mechanism, together with the bacterium's motility and ability to adhere to the gastric mucosa, provides for its survival in the acidic environment of the stomach. *H. pylori* also colonizes the mucosal lining of the stomach.

H. pylori is generally acquired during childhood and can persist for the lifetime of the host. The presence and the role of *H. pylori* has been implicated in a number of gastrointestinal disorders and imbalances. Infection by *H. pylori* may lead to an inflammatory response, increased secretion of gastric acid, and type-B gastritis (Zarrilli et al. 1999; Viganò et al. 2018). Given the normal barrier function of the stomach's mucosal lining, infection by *H. pylori* often leads to a minor concomitant inflammation. In case of chronic manifestation, malignant tumors may result, as has been demonstrated for MALT-lymphoma. The classical anti-*H. pylori* protocol consists of a proton pump inhibitor (PPI) and two antimicrobial agents. The Maastricht/Florence Consensus report, which outlines the diagnostic guidelines and treatment strategies for those with *H. pylori* (Malfertheiner et al. 2017) advises individuals with certain risk factors to undergo eradication therapy. In particular, it is recommended that those with functional dyspepsia, undergo the "test and treat" strategy. However, there remains a lack of options for persons who are either asymptomatic or experience only mild gastrointestinal symptoms. Alternative anti-*H. pylori* protocols are being searched for.

Probiotics in Helicobacter Therapy

The use of probiotics, as monotherapy or synergistic therapy (in combination with antibiotics), is being researched as an alternative way of controlling *H. pylori* infection and reducing side effects of antibiotic treatment (Lesbros-Pantoflickova et al. 2007; Wang et al. 2017b).

Mechanisms by which probiotics work in this application include suppressive effects against gastrointestinal inflammation and against *H. pylori* (Khoder et al. 2016).

Several potential mechanisms by which probiotics may influence *H. pylori* have been discussed. Non-immunological barriers represent the first line of defense against pathogenic bacteria. Probiotics may strengthen this barrier by producing antimicrobial substances, stimulating mucin production, and stabilizing the gut mucosal barrier. Antimicrobial activity of probiotics could be due not only to a direct effect on *H. pylori* but also to the inhibition of its urease activity. In addition, probiotics might enhance the production of prostaglandins, growth factors and anti-inflammatory cytokines (Cain and Karpa 2011). Another possible mechanism is displacement of *H. pylori* through competitive binding to adhesion receptors of *H. pylori* (Mukai et al. 2002).

Previous studies on the effect of probiotics on *H. pylori* have been done with a broad range of probiotic strains on diverse endpoints. A study applying *L. casei* reported a non-significant suppressive effect seen as decrease of urease activity (Cats et al. 2003). Using live cells of a *L. brevis* strain, a reduction of UBT values has been reported, and a correlation to decreased polyamine synthesis was proposed (Linsalata et al. 2004). In case of *L. johnsonii* La1, consumption of whey-based supernatant of *L. johnsonii* La1 culture was found to induce a marked decrease in Helicobacter in infected subjects as measured by breath test analysis, either in combination with Omeprazol or alone (Michetti et al. 1999). Subsequent studies showed that La1 given in a fermented milk beverage reduced *H. pylori* load in adults and in children as measured by the UBT (Cruchet et al. 2003). The species *L. reuteri* has been named in several studies to be active in *Helicobacter* management and in reducing side effects of standard treatments (Lionetti et al. 2006; Francavilla et al. 2008; Scaccianoce et al. 2008).

Emara et al. (2014) used a *L. reuteri* preparation (a mixture of strains *L. reuteri* DSM17938 and *L. reuteri* ATCC PTA6475 in combination with a triple therapy. The *Lactobacillus* supplementation increased the Gastrointestinal Symptom Rating Scale (GSRS) score significantly, but did not improve *H. pylori* eradication rate.

However, successful in vivo studies demonstrating the effects of probiotics on *H. pylori* gastritis are limited. This could be due to the adverse physiological conditions of the stomach, such as an acidic environment, gastric enzymes, bile acids and mechanical stress that reduce the survival and metabolic activity of probiotics.

The Helinorm/Pylopass Strategy

Reducing the amount of *H. pylori* in the stomach by selective bacterial-bacterial cell interaction has been proposed as an effective and novel strategy for combating this stomach bacterium.

While *H. pylori* resides in the mucus where it is present in its motile form, mucus is constantly produced by the epithelium and shed into the stomach lumen. This continuously releases planktonic *H. pylori* cells into the stomach. The probiotic strain used in the Helinorm formulation, *L. reuteri* DSM 17648 (Pylopass), specifically captures such *H. pylori* cells. Pylopass has been carefully selected to enable a rapid and sensitive and highly selective trapping of *Helicobacter pylori* cells in the stomach lumen. Once bound, co-aggregates are flushed from the stomach by natural bowel movement. *L. reuteri* DSM17648 has been developed to specifically co-aggregate with *Helicobacter pylori* under gastric conditions.

Reducing the amount of *H. pylori* in the stomach by selective bacterial-bacterial surface interaction represents an alternative method for managing the stomach pathogen.

The Pylopass Strain

The strain *L. reuteri* DSM17648 was selected by laboratory screening assays among *Lactobacilli* strains of a large culture collection (ORGANOBALANCE GmbH, Berlin, Germany).

Among 700 *Lactobacillus* strains, only 8 were found to co-aggregate with spiral forms of *H. pylori* DSM21031. One of these strains — *L. reuteri* DSM17648 — was analysed in depth and used for product development.

Fluorescence images of co-aggregates of cells stained with HI or CFDA (Fig. 1), then allowed to co-aggregate (Fig. 1B), demonstrated that both DSM 17648 and *H. pylori* participated in the aggregation but did not auto-aggregate.

Quantification of co-aggregate formation between *L. reuteri* DSM17648 and *H. pylori* DSM21031 by flow cytometry revealed that one *Lactobacillus* cell binds 2–3 *Helicobacter* cells.

Co-aggregates can be seen visually as flocking structures, whereas no such structures are present in controls (Fig. 2). Co-aggregation between *H. pylori* and *L. reuteri* DSM17648 occurs within seconds.

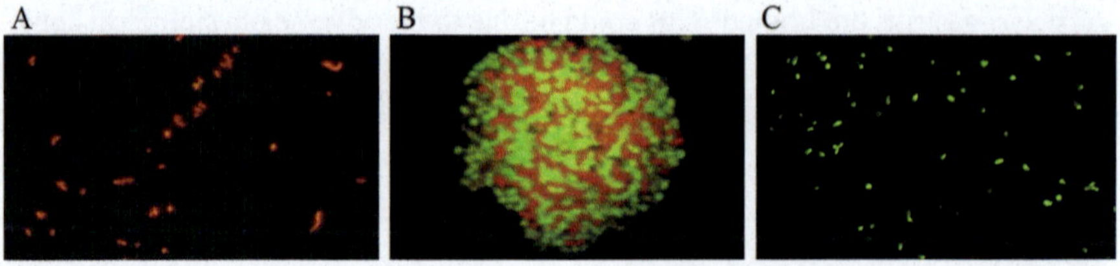

Fig. 1 Co-aggregation between *Helicobacter pylori* DSM 21031 and *Lactobacillus reuteri* DSM 17648 incubated in artificial stomach juice (pH 4). Phase contrast microscopy with fluorescence filter. Magnification 1000×. *H. pylori* was stained with hexidium iodide (**A**), *L. reuteri* was stained with CFDA (**C**). Co-aggregate shows clumping of both strains (**B**). To confirm that both species were present in the aggregates, cells were stained separately using either hexidium iodide or carboxyfluorescein diacetate succinimidyl ester. Both DSM17648 and *H. pylori* DSM21031 participate in the aggregation

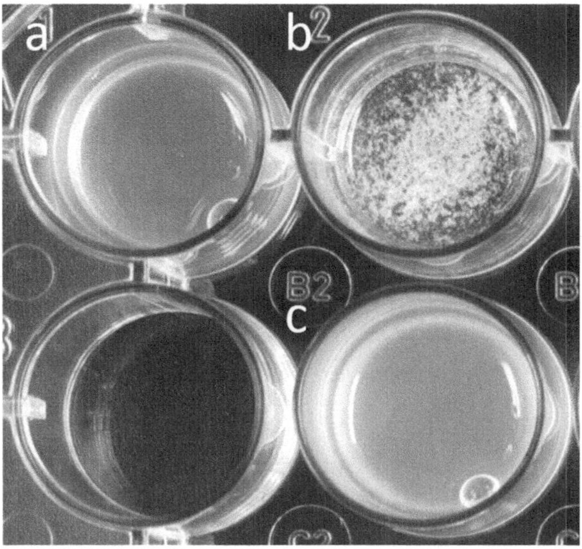

Fig. 2 Co-aggregation is macroscopically visible. Co-aggregation of *Helicobacter pylori* DSM 21031 T and *L. reuteri* DSM 17648 incubated in artificial stomach juice pH 4/ PBS (B). No co-aggregation was observed when single strains were analysed (A: *H. pylori* only, C: *L. reuteri* DSM 17648 only)

Pylopass specifically co-aggregates with *H. pylori* under in vitro conditions and in artificial gastric juice, but not with other bacteria of the commensal intestinal flora.

Numerous other *L. reuteri* strains were tested in the screening process for co-aggregation with *H. pylori*. None of them formed co-aggregates with *H. pylori*, underlining that the effect exhibited by Pylopass is unique and specific.

This specific binding masks the surface structures of *H. pylori* and interferes with *Helicobacter* motility. The aggregated pathogens no longer adhere to the gastric mucosa and are flushed out of the stomach.

Characteristics of Pylopass

Expression of co-aggregation activity is dependent on the growth phase of *L. reuteri* DSM17648, and it is manifest at entry into stationary growth and during stationary phase (Fig. 3). It is dependent on a *L. reuteri* DSM17648 cell surface factor that is present at the end of the exponential growth phase and during stationary phase.

Co-aggregation takes place in the presence of sugar (sucrose, lactose, glucose, fructose, maltose, isomaltose, sorbitol). It occurs at comparable efficiency at room temperature and at 37 °C, and co-aggregation activity is observed over a wide pH range (pH 2.0 – corresponding to empty stomach conditions - up to pH 8, including typical pH values after meals). No pure cultures evidenced auto-aggregation within this pH range. Slightly smaller co-aggregates are formed at pH 2 compared to pH 8 in vitro. Thus, aggregation of *H. pylori* by *L. reuteri* DSM17648 occurs at pH values and conditions encountered in the human stomach.

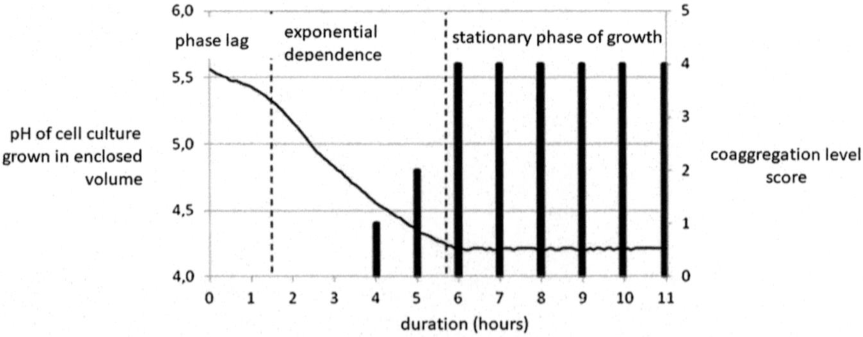

Fig. 3 Relationship between growth phase and co-aggregation activity of *Lactobacillus reuteri* DSM 17648

To understand if proteins are involved in binding, the susceptibility to protease inactivation of the co-aggregation determinants on the surfaces of both DSM17648 and of *H. pylori* was tested after treatment with protease *Strep. griseus* Type XIV, proteinase K, trypsin, or pepsin. Incubating *L. reuteri* DSM17648 with any protease before co-aggregation reduced binding to *H. pylori* DSM21031 by 30%, but did not eliminate it completely. H. pylori required pretreatment with the protease pepsin (as is naturally present in gastric fluids) to be fully active in co-aggregation with *L. reuteri* DSM17648.

Cell Walls Exhibit the Co-Aggregation Sites

SEM images of co-aggregates were prepared to analyse cellular sites of the attachment (Fig. 4).

Clearly, single cells of *L. reuteri* DSM 17648 bind several *H. pylori* cells resulting in a cross-linking of the co-aggregates; and the binding sites on DSM 17648 appear evenly distributed over its surface, binding sites on *H. pylori* do not seem to be exposed on the flagella.

The Power of Metabiotic Formulation

Metabiotics are 'non-viable' bacterial products or metabolic byproducts from probiotic microorganisms that have biological activity in the host. Interestingly, no viable cells of Pylopass are required for the specific action towards *H. pylori*. The co-aggregation activity is preserved during lyophilisation or spray-drying of whole cells of *L. reuteri* DSM17648 and persists in non-viable cells (Table 1). Spray-dried or lyophilised *L. reuteri* DSM17648 cells induce co-aggregate formation with the same sensitivity as untreated cells.

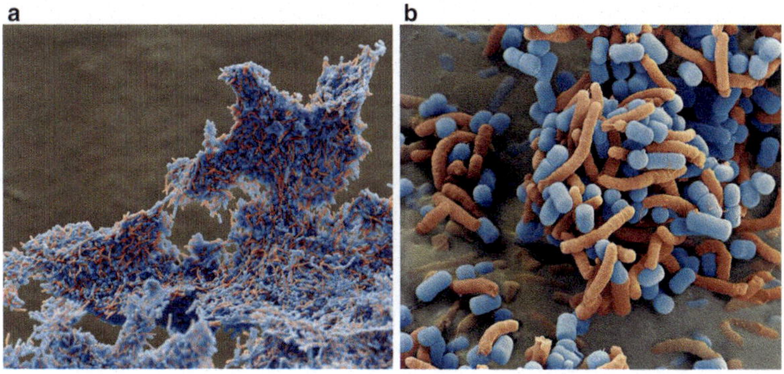

Fig. 4 (**a**) Scanning electron microscopy of co-aggregates consisting of *L. reuteri* DSM17648 (blue) and *Helicobacter pylori* (red), 1.800× magnification (**b**) Scanning electron microscopy of co-aggregates consisting of *L. reuteri* DSM17648 (blue) and *Helicobacter pylori* (red), 11.000× magnification

Table 1 Co-aggregation activity is present in non-viable cells of *L. reuteri* DSM 17648

Values	culture	lyophilised cells	non-viable cells*
Colony forming units [ml^{-1} or mg^{-1}]	$2{,}0 \times 10^9$	$3{,}0 \times 10^7$	0
Co-aggregation score	+++	+++	+++

*Cells were spray-dried and subsequently treated at 40 °C for 24 h

It can be assumed that binding is due to specific surface molecules on the *L. reuteri* DSM17648 cells which are strain-specific and are resistant to such process steps. Such surface molecules might include lipoteichoic acid and carbohydrate structures.

L. reuteri DSM17648 is active as non-viable cell preparation, which greatly reduces any potential side effects of living cultures and helps ensure stable activity in consumer products such as supplements and in medicinal formulations.

Clinical Evaluation

It is hypothesized that *Lactobacillus reuteri* DSM17648 interferes with mobility of *H. pylori* and its adherence to the gastric mucosa by entangling *H. pylori* cells into aggregates and masking *H. pylori* surface sites which are ordinarily available for binding to human gastric epithelium. Once bound, co-aggregates will be flushed from the stomach by natural bowel movement. This novel anti-*H. pylori* activity has not been described previously as a mode-of-action for probiotic treatment of *Helicobacter* infections.

The effectiveness of the metabiotic formulation Pylopass was shown *in vivo*. *Lactobacillus reuteri* DSM17648 (Pylopass) was used in several clinical studies to demonstrate that it effectively reduces the amount of *H. pylori* in the stomach in the absence of antibiotic treatment (Holz et al. 2015; Mehling and Busjahn 2013).

In a study with a placebo controlled design, the effect of using Pylopass for 2 weeks was analysed in a group of persons that were carriers of *H. pylori*. Participants of the study were people aged 18 years or older that had been diagnosed as *H. pylori* positive using the standard 13C urea breath test (UBT, Helicobacter Test INFAI®, Dd ≥ 4‰). The test is based on measuring urease activity of *H. pylori* cells. The test's specificity (98,5%) and sensitivity (97,9%) are well comparable to traditional invasive detection methods, such as endoscopy or biopsies. The breath test measures directly the status of *H. pylori* colonisation and it is well suited to detect reduction or eradication of the bacterium. As a result, the amount and the activity of *H. pylori*-urease is quantified. This reflects the degree of colonisation of the stomach with *H. pylori*. The active ingredient in the study was freeze-fried *L. reuteri* DSM17648 taken orally as a tablet, with 5×10^9 cells each, and a daily dose of 4 tablets. A significant reduction of *H. pylori* load as a result of Pylopass application was detected by following intra-individual changes over time, i.e. in one person during supplementation period. Placebo supplementation did not result in significant changes in UBT values. At the end of the supplementation period, most persons having ingested Pylopass were found to have a significantly reduced *H. pylori* load (Fig. 5).

This protective effect lasted for several weeks after the supplementation period. Clinical studies have also been performed in Ireland (Buckley et al. 2018), Russia (in press), and Pakistan (unpublished) and all of them showed the same effect: *H. pylori*

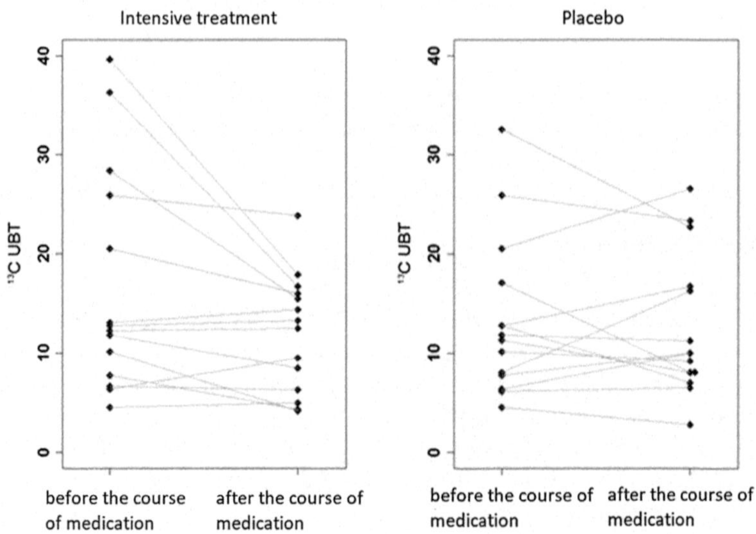

Fig. 5 Absolute 13C UBT values of individuals before and after treatment with verum and placebo

load is significantly reduced and symptoms accompanying *Helicobacter pylori* infection are reduced. Adverse side effects were not observed in any participant.

Outlook

Processing probiotic strains to metabiotic formulations is a potent strategy to provide high quality active ingredients in a stable product. Knowing the mode of action of a product answers the scientific and consumer needs and paves the way for developing highly effective and safe novel bioactives.

Metabolite Metabiotics

Metabolite metabiotics contains in its composition metabolites produced by probiotic strains. We will describe metabolite metabiotics using the example of Bactistatin.

Among the wide range of probiotics used worldwide for treatment and prevention of diseases associated with microecological digestive tract disorders, a prominent role is played by probiotics on the base of live spore-forming gram-positive bacteria belonging to *Bacillus subtilis*. The representatives of this species are among the most ancient microorganisms on our planet; that is why humans and other complex eukaryotic organisms are very well adapted to these commensal microorganisms. Isolated as a pure culture over 100 years ago, these bacilli are extensively represented in samples of soil, water, plant (especially on the base of soybeans) and dairy food products. In healthy people, *B.subtilis* is ever-present in the gastrointestinal tract in the amount of $10^{5–7}$ CFU/g, which is comparable with the quantitative content of lactic acid bacteria (for instance, lactobacilli) in the large intestine. In case of limiting food substrates in culture medium, *B. subtilis* produces spores. Multi-year research had shown that these commensal bacteria are safe for humans, which enabled the regulatory bodies in the USA (FDA), European Union (EFSA), Russia (GosSanEpidNadzor) and in other countries to consider *B. subtilis* strains to be safe (GRAS) for use as probiotic dietary supplements. *B. subtilis* appeal for probiotics production is explained not only by their safety characteristics, but also by their marked biotechnological potential (methodologies of their genetic and molecular taxonomic identification have long since been developed and tested; they have high growth and productivity rates in aerobic and anaerobic conditions in different cheap growth substrates, high rate and broad spectrum of production of enzymes, primary and secondary metabolites, antimicrobial substances, other proteins and low molecular substances; spore production allows for stable and long-term preservation of the genetic characteristics of starter cultures). Different *Bacillus* species are among the most powerful producers of bioactive secondary metabolites (up to 800 compounds) (Ilinskaya et al. 2017). Overproduction of antimicrobial peptides and enzymes and proven biotechnologies are already in widespread commercial use.

The above-mentioned characteristics allowed for considering the strains of these species as "the most perfect multifunctional probiotic bacteria" (Olmos and Paniagua-Michel 2014; Sorokulova 2013; Suva et al. 2016). Broad use of probiotics on the base of live *B. subtilis* strains for people, animals, birds and aquacultures have enabled to set multiple targets and mechanisms of their beneficial effect on live eukaryotic organisms. Bacillar probiotics form SCFA (acetic, propionic, butyric and, less markedly, isobutyrate, isovalerate), synthesize antimicrobial substances that suppress the growth of pathogenic and opportunistic microorganisms, reversing the symptoms of diarrhea, secrete various amylolytic, lipolytic, cellulolytic and proteolytic enzymes which are stable in acidic environment, stimulate the immunity, inhibit intestinal inflammation, normalize the growth of gut microbiota and stimulate the propagation of gut symbiotic lactobacilli and bifidobacteria. The probiotic effect of *B. subtilis* is attributed to a quorum-sensing peptide, that takes part in communication, proliferation and sporulation of these bacteria. Also, it displays anti-inflammatory effects by modulating the synthesis of IL-4, IL-6, IL-10 cytokines and producing the synthesis of anti-shock peptides (Ilinskaya et al. 2017). The research into *B.subtilis* immune stimulating activity have revealed the ability of these bacilli and their spores to sustain gut barrier, to promote inborn and adaptive immunity, activation and development of lymphoid tissue, to stimulate the maturation of antigen-presenting cells (phagocytes with macrophage and dendritic phenotypes), to stimulate cytokine production by macrophages and the formation of IgA, IgG and IFN-γ (Sebastian and Keerthi 2014; Sorokulova 2013; Quintana et al. 2014).

The potential ability of *B.subtilis* to normalize the intestinal inflammation processes are attributed to their capability to inhibit the growth of pathogenic bacteria in the digestive tract and increase the levels of bifidobacteria and lactobacilli (Lefevre et al. 2015; Suva et al. 2016). Antimicrobial agents formed by various *B.subtilis* strains include compounds differing in their chemical composition (various peptides, bacteriocins, antibiotics, lytic enzymes, organic acids; vitamins, folate and biotin in the first place, and others); the spectrum of their effects includes a broad range of microorganisms. Bacillar antimicrobial peptides can form membrane channels and destroy the cell wall of pathogenic and opportunistic microorganisms, including antibiotic-resistant ones. They have antibacterial, antimycotic, antiprotozoal and antiviral action, as well as immunosuppressive and antitumor effects, they can inhibit quorum-sensing and the related virulence and biofilm formation. Also, bacilli produce a lot of ribosomally and nonribosomally connected similar peptides (over 20 lipoproteins, glycopeptides, cyclic peptides, bacteriocins and antibiotics). They are synthesized from almost 300 different precursors. Then they can be modified by methylation, acetylation, glycosylation, formation of heterocyclic rings (Ilinskaya et al. 2017; Perez et al. 2017; Phister et al. 2004; Sorokulova 2013; Sumi et al. 2015; Suva et al. 2016). Bacilli excrete amino acids and vitamins into the culture medium; bacillar probiotics administration reduces the levels of serum cholesterol and triglycerides in model animals.

The *B. subtilis* pentapeptide can activate the key ways to preserve gut epithelial cells. *Bacilli* are involved in the metabolism of food components, xenobiotics and therapeutics, sustaining intestinal homeostasis in healthy people. All the above-mentioned biological characteristics of *B. subtilis* are strain-specific (Ilinskaya et al. 2017; Sorokulova 2013). Moreover, bacilli produce extracellular vesicles, spherically-shaped membrane-like structures, in the composition of which proteins, lipids, nucleic acids and metabolites are found that show biological activity. These vesicles take part in the interaction of these bacteria both with other microorganisms and with the host cells (Kim et al. 2016).

Within a period from 1994 to 2006, by using the *Bacillus subtillis* strain No. 3 (deposited with the All-Russian Collection of Microorganisms, VKPM V-2335), Russian researchers had developed the technology for generation of the Bactistatin dietary supplement which is a germfree filtrate of cultural fluid of the *B. subtilis* strain No. 3 grown by pour plate method in liquid culture medium on the base of soy protein hydrolysate. Bactistatin contains a complex of low molecular weight compounds (total carbon and nitrogen contents are 9.6 g/l and 2.2 g/l, respectively) immobilized on zeolite in admixture with soy flour hydrolysate; it comes in powder-containing capsules. The zeolites of Holinsky deposits are micropore silicate minerals containing ions of sodium, potassium and calcium, easily exchanging among themselves and with the surrounding substrate. The acid hydrolysate of soy flour with maximum protein breakdown level provided high bacilli biomass yields. Bactistatin examination has shown that a liter of freeze-dehydrated supplement contains low molecular weight compounds with antimicrobial activity (such as bacilisin, subtilin, iturin, globicin, mycosubtilin; bacillin, obutin, endosubtilisin), enzymes splitting carbohydrates, fats and proteins (such as nitrate reductase, α-amilase, pyruvate kinase, ribonuclease, aminopeptidase, subtilopeptidase), lysozyme, catalase, polypeptides, peptidoglycans, polysaccharides, soybean components etc. (Volkov 2008; Volkov et al. 2006; 2007; Vorobeichikov et al. 2008).

In subsequent metabolome studies of germfree cultural fluids of this strain of probiotic *Bacillus subtilis* with the use of high-efficiency liquid chromatography; high-efficiency liquid chromatography with tandem mass-selective detection; high-efficiency liquid chromatography with tandem high-resolution mass-selective detection with electrospray ionization; high-efficiency high-resolution liquid chromatography with tandem high-resolution mass-selective detection; gas chromatography; gas chromate-mass-spectrometry — over 100 other low molecular weight metabolites had been tested including B vitamins (B5, B2, B1, B6), C vitamins and PP (nicotinic acid), free amino acids (alanine, arginine, asparagine, aspartic acid, valine, hydroxyproline, histidine, glycine, glutamine, glutamic acid, isoleucine, lysine, methionine, leucine, ornithine, proline, serine, tyrosine, threonine, tryptophan, phenylalanine, cystine, citrulline), short-chain (acetic, butyric, propionic), free (16) and esterified (20) fatty acids, metabolites of aromatic amino acids – tryptophan and polyphenols.

The following acids had the highest levels in cell-free concentrate of *Bacillus subtilis* (strain No. 3) cultural fluid: palmitic, stearic, myristic, palmitoleic, pentadecanoic, lauric, capric, margaric and lignoceric acids. Bactistatin also includes 18 chemical elements (*Na*, *Mg*, *K*, *Ca*, *Mn*, *Fe*, *Co*, *Cu*, *Zn*, *Se*, *Cr*, *Mo* etc.). Additionally, amikacin A and B (dehydroisocoumarin derivatives) have been found in Bactistatin, having antibacterial, antimycotic, antitumor and antiviral effects, phengicins A, B and C (cyclic lipopeptides with antimicrobial activity), methylbutanoic, lactic, hydroxisocaproic, oxalic, phosphoric, phenylacetic, hydrocinnamic, phenyl-lactic, 3-hydroxyphenil-lactic, 4- hydroxyphenil-lactic acids, methacryloyl-glycine, phenol class compounds (phenol, catechol (1,2-dihydroxybenzene, pyrocatechol, hydroquinone, 3-methylcatechol), peptides (DL-alanyl-L-leucine) and cyclo(leucylprolyl) (L-Phe-D-Pro lactam), unidentified cyclic dipeptide (3- benziohexahydropyrrol[1,2-a]pyrazine-1,4-dione (Cyclo(D-phenylalanine-L-prolyl) and other three unidentified cyclic dipeptides, some identified nitrogen compounds (caprolactam, phenylethylamine, phenylalanine, thymine, uracil, dihydrouracil, indole, tyramine, 5-hydroxi-1H-indole, 9H-purin-6-ol, 13-docosenamide, erucic acid amide), alcohols and glycols (ethylene glycol, propylene glycol, methyonol, benzyl alcohol, heptaethylene glycol, nonaethylene glycol, decaethylene glycol). The above mentioned low molecular weight compounds found among Bactistatin components can potentially be involved – independently or in combination with other nutrients and microbial bioactive molecules – in aerobic respiration, in energetic and redox reactions, get embedded within a multitude of biochemical natural compounds, in complex aggregates of lipids, proteins and carbohydrates, in cell membranes, mitochondria and other subcellular structures.

Clinical Assessment

Clinical trials of Bactistatin with various animal models and in humans have shown that oral administration of Bactistatin promotes the intestinal digestion of carbohydrates, lipids and proteins, reduces diarrhea syndrome, has marked immunomodulating and detoxicating effects. Bactistatin prescription normalizes the microbial ecology of the digestive tract, restores gut barrier function, stimulates the growth of symbiotic lactobacilli and bifidobacteria, inhibits the propagation of pathogenic and opportunistic bacteria, fungi and some viruses. Also, Bactistatin restores many other clinical symptoms of gastrointestinal disorders in animals and humans caused by short- and/or long-term exposure of gut symbiotic microbiota to biological and chemical factors and agents (Ardatskaya 2015, Ardatskaya et al. 2015; Volkov 2008; Volkov et al. 2006; 2007; Minushkin et al. 2006; Shenderov et al. 2018a). It is not improbable that multiple low molecular weight compounds of the Bactistatin metabiotic can act as molecules that normalize acid-alkaline balance in cells and tissues, as signals that remodel membranes, mitochondria and other intracellular organelles involved in metabolic processes, in the development of brain cells and endocrine APUD system, as regulators of intercellular ion channels modulation, as epigenetic components that help accomplish gene expression in the metagenome and meta-epigenome of the organism, post-translational protein modification and other biochemical reactions in live organisms that remain not quite clear until now (Oleskin et al. 2016; Shenderov 2017; Shenderov 2018; Shenderov et al. 2018a).

Table 2 New complex metabiotics on the base of *B. subtilis*

Composition	Indications for use
Metabiotic *B. subtilis* + coenzyme Q10 + resveratrol	Health of cardiovascular system, metabolic syndrome
Metabiotic *B. subtilis* + polysaccharide complex (β-glucan and mannan)	Immune boosting
Metabiotic *B. subtilis* + hyaluronic acid + zinc + vitamins C and E	Skin health
Metabiotic *B. subtilis* + taurine + plant hepatoprotector (*Phylantus amarus*)	Liver and hepatobiliary system health
Metabiotic *B. subtilis* + multivitamins + multi-minerals	Complex nutritive support

Modification of this product is aimed at creation of combined therapeutics incorporating targeted active ingredients. This approach allows for creating metabiotics that contribute to targeted health support with respect to particular systems of the organism (Table 2).

Prospects in the Field of Intended-Use Metabiotics Creation

The analysis of contemporary literature bears evidence that the development of therapeutics intended for optimum formation, preservation and restoration of human symbiotic microbiota and their practical implementation are among the critical ways of preventive and regenerative medicine development. On the one hand, search is ongoing for new application targets of traditional probiotics made on the base of live probiotic strains and their detailed analysis, on the other hand, biologically active microbial molecules or their complexes are being isolated and identified which determine their application targets, efficiency and possible negative consequences of using these probiotics. New potential probiotic strains capable of modifying microelement, antioxidant, anti-inflammatory, psychic statuses, epigenotype, quorum-sensing signaling are selected among both dominant microorganism species and among species represented on the skin and mucosa in low amounts (Shenderov 2014b). Supposedly, knowing the range and concentration of low molecular weight biologically active compounds formed by known or potential probiotic strains will help construct new effective metabiotics and set their optimum frequency, doses and method of administration. The selected strains of human origin with unique functional or biochemical activity can become perspective starter cultures for large-scale biotechnological production of microbial bioactive special-purpose substances. Their use as metabiotics in their own right or as fortifiers of known traditional probiotics, dietary supplements and functional foods will help develop and implement more efficient microecological products for symbiotic microbiota reservation and restoration in the majority of the population, in specific population groups and in individuals. It should also be remembered that beneficial effects of the prospective metabiotics depend to a great degree on the composition of the growth medium, time and conditions for initial starter culture incubation, methods of lysis of microbial cells, isolation, purification and (optionally) encapsulation of the corresponding low molecular weight components of a microbial cell, their metabolites and signaling molecules.

More intensive and detailed research is needed into the genome, epigenome and metabolome of known and potential probiotic strains in order to identify the genes

B. A. Shenderov et al., *METABIOTICS*,
https://doi.org/10.1007/978-3-030-34167-1_15

associated with the synthesis of particular biologically active compounds. It will allow for developing improved technologies for production of microbial compounds with increased biological and pharmacological activity and a marked technological, financial and economic appeal. Practical implementation of the "metabiotics" concept will give an option of supplying biotechnology not only with strains belonging to traditional probiotic species (bifidobacteria, lactobacilli, streptococci, leuconostocs, pediococci, propioni bacteria, *E. coli*, enterococci, bacilli, saccharomycetes and other rarer species of probiotic microorganisms), but also with dozens and hundreds of strains belonging to other species of dominating phyla of symbiotic microorganisms (*Bacteroides, Firmicutes, Proteobacteria, Actinobacteria, Archae*) (Shenderov 2017; Shenderov et al. 2018a; Engevik and Versalovic 2017). Broadening the spectrum of low molecular weight microbial components and metabolites (antimicrobial agents, regulators of energy metabolism in the host's mitochondria, compounds that reduce triglyceride levels, low-density lipoprotein cholesterol, immune response modifiers, antidepressants, memory boosters etc.) found in symbiotic microorganisms that dominate in the human organism ecosystem is conductive to a sharp increase in the range of produced metabiotics. They can become safer and more effective than those already created on the base of bioactive compounds synthesized by traditional probiotic strains of bifidobacteria, lactobacilli, *E. coli*, bacilli and yeasts (Dorrestein et al. 2014; Shenderov 2013c; Sonnenburg and Fischbach 2011; Sonnenburg and Backhed 2016; Wikoff et al. 2009).

Establishing databases of individual low molecular weight biologically and pharmacologically active compounds or their groups formed by representatives of various species of human symbiotic microorganisms can play a key role in the creation of new intended-use metabiotics, and subsequently — synthetic and semi-synthetic metabiotics. The example of such synthetic metabiotic is Actoflor C (the complex of 12 organic and amino acids) produced in Russia (Vakhitov and Sitkin 2014). In the recent years, new gene clusters have been found in the representatives of gut microbiota, which are responsible for the production of new specialized metabolites (such as polyketides, sterols and isoprenoids, nonribosomally synthesized peptides etc.), whose functions are almost completely unknown (Dorrestein et al. 2014). Expanding our knowledge in microbial metabolites and signaling molecules contained in biological fluids of healthy people belonging to different ethnic groups may also be of great diagnostic value and can become a pre-requisite for the construction of express test systems for metabolome assessment in healthy and ill people (Beloborodova and Osipov 2000; Engevik and Versalovic 2017; Gilbert et al. 2016; Holmes et al. 2011; Li et al. 2008; Nicholson et al. 2005; Nicholson et al. 2012; Shenderov 2013a).

Some New Targets and Approaches to the Construction of Intended-Use Metabiotics

Tables 1, 2, 3, 4, 5, 6, 7 and 8 show some new targets and approaches to the use of low molecular microbial molecules for the development and production of essentially new metabiotics in order to prevent and treat common conditions associated with microecological imbalance in humans. These approaches, if realized, will enable to promptly and efficiently use the potential of the basic representatives of human symbiotic microbiota when creating targeted metabiotics.

Modulators of QS-Regulation

Quorum sensing (QS) is a cell-to cell communication allowing bacteria to sensing small diffusible signaling molecules. QS microbial intercellular signaling coordinate gene regulation in signal bacterium species within a community as well as across kingdoms and controls various types of cooperative social behaviours (biofilm formation, virulence traits, metabolic demands, antibiotic resistance, host-microbe interactions etc). In the long term, these microbial bioactive molecules are likely to become – individually or in combination – the base for potential QS metabiotics (Table 2).

Modulators of Immunity

Probiotic microorganisms actively and in various ways affect local and systemic immunity, participate in maintaining colonization resistance of the skin and mucous membranes, induce oral tolerance, reduce the frequency and risk of allergic disorders, inhibit the growth of pathogenic and opportunistic bacteria, produce various antibiotics, bacteriocins and other antimicrobial compounds, increase the activity of phagocytic cells and natural killer cells, stimulate the production of IgA, inhibit the

B. A. Shenderov et al., *METABIOTICS*,
https://doi.org/10.1007/978-3-030-34167-1_16

Table 1 Some approaches to the construction of intended-use metabiotics (Shenderov 2017; Shenderov et al. 2018a; Shenderov 2011a; Shenderov 2011b)

Item No.	Approaches
1	Modulators of *quorum-sensing* regulation (QS-metabiotics)
2	Modulators of immune (immuno-metabiotics), antioxidant (antioxi-metabiotics), neuropsychic (psycho-metabiotics) statuses
3	Modulators of energy metabolism in mitochondria and gut microbiota (energy-metabiotics)
4	Modulators of epigenome regulation of phenotypic gene expression and post-translational effects (epigeno-metabiotics)
5	Modulators of intracellular information exchange in the populations of prokaryotic, eukaryotic cells and between the host cells and bacteria (info-metabiotics)
6	Modulators sustaining the genome and microbiome stability, preventing large intestine and skin neoplasms and other targets

Table 2 Some microbial low molecular weight compounds capable of modulating *quorum-sensing* regulation of microbial communities, cells of eukaryotic tissues, and probably sub-cell organelles (Shenderov et al. 2018a; Engevik and Versalovic 2017; Shenderov 2011a; Shenderov 2011b; Thompson et al. 2015; Singh et al. 2018; Verbeke et al. 2017)

Item No.	Compounds
1	Inhibitors of protein synthesis (for example, antibiotics that suppress protein synthesis on ribosomal level; different microbial peptides)
2	Receptor-ligand relationship antagonists (for example, microbial fatty acids trans-isomers; bacteriocins)
3	Inhibitors of acyl-HSL signaling (for example, microbial halogenated furanones)
4	Inhibitors of histidine kinase
5	Enzymes that degrade QS-autoinducers (for example, microbial acylases, lactonases, specific proteases similar to serpins of bifidobacteria)
6	Synthetic analogues of microbial autoinducers imitating signaling molecules
7	Micronutrients of microbial, herbal or animal origin, interfering QS regulation (for example, peptides, lactones, lectins, polyphenols)

proliferation of lymphocytes, modifies the profile and the number of synthesized cytokinins/chemokinins, affect the differentiation and maturation of dendrid cells, stimulate and change the nature of Th1 ITH2 responses, synthesize the production of vitamins B and K, exopolysaccharides, bifidogenic factors, conjugated linoleic acid, regulate the synthesis of various intestinal epithelial cells antimicrobial peptides (defensins) and other substances involved in the implementation of immune effects. Low-molecular metabolites, surface and deep structures of probiotic bacterial strains (Table 3) have specific receptors and targets in various parts and cells of the immune system.

Table 3 Low molecular weight antimicrobial compounds and effector structural components related to probiotic (symbiotic) microorganisms, as the base for potential immuno-metabiotics (Chervinets et al. 2018; Shenderov et al. 2018a; Blander et al. 2017; Engevik and Versalovic 2017; Kamada and Nunez 2014; Karczewski et al. 2014; Laino et al. 2016; Lebeer et al. 2018; Proal et al. 2017; Singh et al. 2018; Zhernakova et al. 2016)

Item No.	Compounds
1	Antimicrobial effect (lactic, acetic, propionic, butyric, benzoic, other organic acids, hydrogen peroxide, carbon dioxide, nitrogen oxide, diacetyl, bacteriocins, microcins, bacteriocin-like antibiotics, defensin-like peptides, enzymes with antimicrobial effect (lysozyme), biosurfactants, polyamines, lectins etc.)
2	Structural components (surface S-proteins of fimbriae, peptidoglycans, lipoteichoic acids, exopolysaccharides, LPS, nucleic acids etc.) and microbial cell metabolites (various peptides, proteins, DNA, rich in CpG loci, SCFA, homoserine lactones, dopamine, serotonin, metabolites of histamine and tryptophan etc.) of probiotic microorganisms that interact with specific receptors and targets in different components and cells of the immune system.

Table 4 Modulators of energy metabolism in mitochondria and gut microbiota as a base for potential energy-metabiotics (Shenderov 2018; Engevik and Versalovic 2017; Sharma and Shukla 2016; Wallace 2010; Wallace and Redinbo 2013; Singh et al. 2018)

Item No.	Compounds
1	Vitamins (B1, B2, B3, B5, B6, B7, B9, B12, C, K1, K2)
2	Non-vitamin substrates (NAD, NADH, NADP+, NADPH, ATF, cytidine triphosphate, S-adenosyl methionine, 3'phosphoadenosine-5′-phosphosulphate, glutathione, coenzyme B, coenzyme M, coenzyme Q10, co-factor F-430, heme, alpha lipoic acid, methanofuran, molibdopterin/molybdenum co-factor, pyrroloquinoline quinone, tetrahydrobiopterin, tetrahydromethanopterin)
3	Minerals (*Ca*, *Cu*, Fe^{++}, Fe^{+++}, *Mg*, *Mn*, *Mo*, *Ni*, *Se*, *Zn*)
4	Amino acids (arginine, lysin, methionine, cysteine, β-alanine, serine, threonine, histidine, tryptophan, aspartic acid, carnitine)
5	Organic acids, Krebs cycle participants
6	Some nucleotides (for example, pyrimidine), microRNA etc.
7	Combinations of the listed low molecular weight microbial and herbal compounds

Modulators of Energy Metabolism

Energy-related processes in mitochondria and their functional analogues (inner membranes) in bacteria need a multitude of enzymes and co-factors (Table 4).

Table 5 Antioxidant activity of low molecular microbial compounds as a base for creating potential antioxi-metabiotics (Shenderov et al. 2018a; Engevik and Versalovic 2017; Minushkin et al. 2006; Jones and Neish 2017; Mishra et al. 2015; Prosekov et al. 2015; Wang et al. 2017a)

Item No.	Mechanisms of antioxidant activity
1	Binding of metal ions involved in the host's oxidative reactions (for example, Fe^{2+}, Cu^{2+})
2	Microbial antioxidant enzymes (superoxide dismutases-Fe-SOD, Mn-SOD; catalase etc.)
3	Microbial structural components and metabolites with antioxidant effect (proteins, peptides, polysaccharides, glutathione, butyric acid, folate, vitamin B12, thiamine), reducing oxidative stress due to directly or indirectly increased synthesis of the host antioxidative enzymes
4	Microbial compounds involved in the regulation of host signaling antioxidant pathways (Nrf2-Keap1-ARER; NF-κB, protein kinase MAPK, PKS) engaged in maintaining redox-potential of the organism
5	Microbial components modulating the host cell regulation of oxidative free radicals synthesis
6	Microbial compounds (for example, organic acids, bacteriocins, biosurfactants) that restore the host gut microbiota and that, on the contrary, inhibit proliferation of pathogenic and opportunistic microorganisms and the related intestinal endotoxemia, metabolic disorders and oxidative stress

Modulators of Antioxidant Status

Oxidant/antioxidant system regulates the processes of free radical oxidation. The main active forms of oxygen (ROS), capable of damaging the membranes, nucleic acids and other vital cell structures, are superoxide radicals, hydrogen peroxide, hydroxyl (free) radicals, singlet oxygen forms, ions NO2-. The level of ROS formed in the body during various physiological processes is controlled by a multistage system of endogenous and exogenous bioantioxidants, which support free radical oxidative processes in physiological conditions and protect cell structures from destruction. Imbalance between oxidation reactions associated with ROS and reactions of their neutralization leads to uncontrolled increase in ROS concentration and development of oxidative stress, which leads, first of all, to disruption of the structure and functions of the membranes of various cells. Normally, the content of ROS in the body is effectively regulated by a system of antiradical and antiperoxide protection, which includes both enzyme and non-enzyme systems. Symbiotic microbiota is also actively involved in the fight against oxidative stress of various origins. Oxidative stress and imbalance of antioxidant defense mechanisms (the amount and activity of enzyme and non-enzyme antioxidants and antioxidant systems of symbiotic microbiota) play an important role in the pathogenesis of many metabolic diseases (chronic fatigue syndrome, metabolic syndrome, premature aging and others). To reduce the risk of such diseases, it is recommended to additionally introduce microbial and food antioxidants in the form of various functional foods (Table 5).

Table 6 Neuromodulation effects and targets of microbial ingredients as a base for potential psycho-metabiotics (Oleskin and Shenderov 2016; Shenderov 2016a; Shenderov et al. 2018a; Engevik and Versalovic 2017; El Aidy et al. 2016; Lebeer et al. 2018; Maguire and Maguire 2019; Oleskin et al. 2017; Singh et al. 2018)

Item No.	Effects and targets
1	Trimethylamine is involved in social communication (attractive smell)
2	Bile acids. Dysbioses accompanied by bile acid metabolism disorders caused by gut microorganisms, often result in the risk of hepatic encephalopathy
3	Amino acids (aspartate, glutamate, glycine, taurine, tryptophan)
4	Phenol and phenol derivatives. Their increased concentrations in urine samples are found in children with autism and schizophrenic patients. They also inhibit dopamine transformation into noradrenaline
5	Indole is a microbial metabolite formed from tryptophan. Indolepropionic acid protects neurons from oxidative damage and death caused by beta amyloid protein effects in people with Alzheimer's disease
6	Tryptamine is a neuropeptide stimulating the release of serotonin. Low levels of tryptamine are found in patients with severe depression
7	Vitamins B12, K, biotin, cobalamin, folates, nicotinic and pantothenic acids, pyridoxine, riboflavin, thiamine are co-factors of metabolic reactions that take part in the synthesis of neurotransmitters and the realization of neurohormones or neuropsychic responses and functions
8	Short-chain fatty acids (SCFA). Propionate influences social behavior, changes brain cell phospholipid composition, as well as neuropeptide levels (serotonin, glutamate, dopamine). Butyrate, gamma-aminobutyrate or gamma-aminobutyric acid (GABA) influence energy homeostasis, nerve cell functions, mood and behavior. Acetate is the basic substrate for the synthesis of acetyl-CoA, it enhances histone acetylation, influences long-term memory, consolidation and neuroprtotection/regeneration. SCFA modify the number of methyl group donors by interfering with the metabolism of monocarbon compounds thus potentially influencing DNA and histone methylation
9	Spermine is a polyamine that slows down age-related memory impairment
10	Microbial hormones and transmitters (serotonin, C-dihydroxyphenylalanine — DOPA, dopamine/noradrenaline, acetylcholine, histamine, tryptamine), acetylcholine, catecholamines, serotonin, histamine and others involved in в the formation of a pool of these compounds and stimulating epithelial, gut enteroendocrine cells and the synthesis of peptide YY, neuropeptide Y, cholecystokinin, glucagon-like peptide-1,2 and substance P in the human organism

Modulators of Intercellular Information Exchange

The relationship between the host and its microbiota is regulated by constant mutual exchange of signaling information (Oleskin et al. 2016; Atkinson and William 2009; Engevik and Versalovic 2017; Schauder and Bassler 2001) which also includes the production by symbiotic microorganisms of SCFA, nitrogen oxide and other gas microbial molecules, glutamate, beta-alanine, biotin, surface proteins, peptides, polysaccharides, lectins, peptidoglycans, lipoteichoic acid, glycopeptides, flagellins and their complexes, lipopolysaccharides and other low molecular weight compounds that

Table 7 Modulation of epigenetic regulation of host gene expression as a target for selecting potential epigeno-metabiotics (Shenderov et al. 2018a; Bhat and Kapila 2017; Feil and Fraga 2012; Roda et al. 2007; Shenderov 2016b; Singh et al. 2018; Wallace 2010)

Item No.	Mechanisms of epigenetic regulation
1	Microbial substrates, co-factors interfering with epigenetic regulation (enzymes, organic acids and amino acids, vitamins, lectins, other coenzymes)
2	Microbial inhibitors and activators of general-purpose epigenetic signaling (volatile fatty acids, gaseous molecules, lectins)
3	Microbial activators and inhibitors having a specific effect on particular enzymatic participants of epigenetic machinery (methyltransferases, demethylases, acetyltransferases, sirtuins, ribosyltransferases, hydrolases, phosphotransferases, kinases, BirA ligase, synthetases, nucleases, DNA and RNA ligases) — Butyrate, microbial derivatives of plant polyphenols
4	Microbial antagonists of receptor-ligand relationship (fatty acid trans isomers, carbohydrate L-isomers, amino acid D-isomers, lectins)
5	Microbial proteases that degrade enzymes, effectors or receptors involved in epigenetic processes (acylases, lactonases, bifidobacterial serpins)
6	Semi- and fully synthetic analogues of different modifiers of epigenetic mechanisms

Table 8 Best-known natural bioactive compounds of dietary or microbial origin involved in epigenomic processes (Shenderov 2013a; Shenderov 2013c; Shenderov et al. 2018a; Bhat and Kapila 2017; Cyr et al. 2013; Feil and Fraga 2012; Jimenez-Chillaron et al. 2012; Li and Tollefsbol 2010; Mischke and Plosch 2016; Remely et al. 2015; Said 2013; Shenderov 2016b; Shenderov and Aleshkin 2012)

Item No.	Compounds
1	Mono-, di-, oligo- and polysaccharides
2	DNA, RNA, adenine, cytosine, guanine, NAD, ATP
3	Polyphenols and carotenoids (epicatechin, genistein, quercetin, hesperidin, luteolin, garcinol, isocyanates, curcumin, resveratrol, coumarin)
4	Acetyl-coenzyme A, coenzyme Q10, S-adenosyl methionine
5	Melatonin, glutathione
6	Acetaldehyde, ethyl alcohol
7	Selenium, magnesium, potassium, zinc, iodine, cobalt, iron, calcium, manganese, copper
8	Pyruvate, citrate, lactate, α-ketoglutarate, succinate, formate
9	Spermidine
10	Folate, B1, B2, B12, B6, C, E, D3, biotin, pantothenate, nicotinic, orotic acids, choline, alpha-lipoic acid
11	Proteins, peptides, arginine, lysine, methionine, cysteine, α-alanine, valine, leucine, serine, threonine, glycine, histidine, tryptophan, aspartic acid, glutamate, acetyl-L-carnitine, carnosine
12	Phospholipids, EPA, DHA
13	Glucosamine
14	Butyrate, propionate, acetate, caprylic acid
15	Sulforaphane, cysteines of cruciferous vegetables
16	Betaine
17	Allyl-mercaptans of garlic

can modify the growth, regeneration and apoptosis of gut epithelial and immune cells, the motility, fermentative activity and structure of intestinal mucosa, the structure of Toll- and other receptors, induce the synthesis of regulatory signals, antimicrobial peptides, enhance or inhibit gut vascularization, receptor activity in the enteric nervous system, to prevent and restore mucosal lesions, gut barrier permeability (Shenderov 2015a; Corthesy et al. 2007; Lebeer et al. 2008; Maguire and Maguire 2019; Marco et al. 2006; Ng et al. 2009; Shenderov 2011a; Shenderov 2013c). The presented information allows for saying that there is a prospective possibility of creating one more line of metabiotic therapeutics – "meta-informobiotics."

Modulators of Neuropsychic and Social Behavior

Over recent years, quite a few publications have appeared demonstrating that certain low molecular weight compounds formed by symbiotic (probiotic) microorganisms can modify behavioral responses in humans, animals and insects (Engevik and Versalovic 2017; Foster et al. 2016; Maguire and Maguire 2019; Oleskin et al. 2017). It is known that over 400 microbial metabolites pass from the gut to the brain under normal and pathological conditions (Chernevskaya and Beloborodova 2018). The obtained knowledge opens up the perspectives of effective microbial products used as meta-psychobiotics both in clinical studies and in medicine and veterinary (Oleskin et al. 2016; Shenderov 2015b; Oleskin and Shenderov 2016). For production of meta-psychobiotics in the nearest future, it is recommended to use symbiotic (probiotic) bacterial strains that form low molecular weight substances with certain neurotransmitter activity or its whole complex (Table 6).

Modulators of Epigenome Regulation

Previously, it was thought that human health and chronic metabolic diseases are predominantly driven by acquired genetic mutations. From the modern point of view, the genetic system should be further supplemented with an epigenetic system responsible for switching genes on or off in response to various endogenous and exogenous factors and agents. Epigenetic changes are inherited changes that are not related to changes in the sequence of nucleotides in the DNA. It is assumed that epimutations occur 100 times more often than genetic mutations. These epigenetic changes are both adaptive and risk factors for various diseases. It has recently been proposal that exogenous, commensal and symbiotic microorganisms can participate in various epigenetic mechanisms relevant to human health and disease. Owing to low molecular weight microbial structural, metabolic and signalic molecules are able to sense environmental factors, interact with corresponding cell surface, membrane, cytoplasm and nucleic acid receptors and modify epigenomic regulation of gene expression and/or alterate the information exchange in numerous bacterial and bacteria-host systems. The design of microbial epigenetic-based remedies and functional foods (prepared on the base of probiotic living bacteria, their structural compounds, metabolites and/or signal molecules) could be a very important scientific-applied approach for the treatment and prophylaxis of epigenetic-associated human diseases. Potential reversibility of epigenetic changes allows for developing metabiotics (Bhat and Kapila 2017; Feil and Fraga 2012; Remely et al. 2015; Shenderov and Aleshkin 2012) which can activate or inhibit the corresponding epigenetic processes and signal transduction (Tables 6 and 7).

Low molecular weight compounds of microbial origin have different effects on epigenetic mechanisms involved in the regulation of gene expression with regard to genes responsible for the formation and activity of various enzymes and other molecules, their absorption, expansion, metabolism and excretion, for the functioning of mitochondria, ribosomes, signaling molecules, ion channels and other molecular formations in the host eukaryotic and microbial cells (Bhat and Kapila 2017; Shenderov 2016b). Moreover, in the process of degradation and metabolization of various dietary components, gut microbiota is capable of endogenous formation of epigenetically active compounds that modulate the work of cellular epigenetic machinery. For example, secondary metabolites of plant polyphenols having aromatic rings and one or many hydroxyl groups and found in vegetables, fruits, green tea, red wine.

Plant polyphenols play an important part in the life of plants protecting them from photosynthetic stress, various reactive oxidants. Over 80% of natural polyphenols undergo microbial degradation with the formation of metabolites that differ in their activity, including the modulation of epigenetic cell processes. The resulting secondary metabolites not only show antioxidant properties in patients with cancer, neurodegenerative diseases, chronic inflammatory bowel disorders, but also directly influence enzymes, other proteins, receptors and signaling cell pathways. It was found that beneficial effects of such secondary metabolites on mammal organisms is realized through the modulation of NF-kB expression, chromatin remodeling by changing the fermentative activity of deacetylases (HDAC) and DNA-methyltransferases (DNMT) involved in epigenomic gene expression. The example of such plant polyphenols that have undergone microbial modification are curcumin, genistein, resveratrol etc. (Russell et al. 2013). Food components (butyrate, sulforaphane, curcumin) act upon histone-acetyltransferase (HAT) and HDAC activities. NAD+, S-adenosyl methionine and 2-oxoglutarate interfere with the processes of DNA and histone methylation, acetylation and ribosylation (Cyr et al. 2013). Water soluble B vitamins (biotin, niacin, pantothenic acid) cause various histone modifications. For example, biotin is a substrate for histone biotinylation, niacin — for histone rybosylation. Resveratrol, butyrate, sulforaphane and diallyl sulfate inhibit HDAC activity, while curcumin inhibits the activity of histone acetyltransferases (Choi and Friso 2010).

Introduction into preventive medicine of active microbial metabiotics – individually or in combination with specially chosen herbal compounds – will help to effectively deal with pathologic conditions resulting from epigenomic or informational disorders (atherosclerosis, diabetes mellitus type 2, hypertensive disease, cancer, Alzheimer's and Parkinson's diseases, other neurodegenerative, metabolic and autoimmune chronic conditions).

Metabiotics on the Base of Microbial Gas Molecules

Components and metabolites of anaerobe and archaebacteria involved in the synthesis and modulation by microbial and eukaryotic cells of various gas molecules (*NO*, *CO*, H_2S, H_2, CH_4, NH_3 etc.) may be viewed as potential metabiotics. Many of these gas molecules represent very ancient signaling molecules. They are formed in the organism by way of microbial transformation of different compounds, displaying their effects via neuroendocrine, immunological and biochemical responses, modulating redox-signaling, regulating the activity of ion channels and ion transporters in different cells, interfering practically with all physiological functions, biochemical reactions and behavior, influencing the stability of metagenome, epigenome, phenotypic gene expression and post-translational modification of cell products. For example, methane can change intestinal motility; constipations which are not infrequently caused by impaired microbial methane production, often precede the development of neurodegenerative diseases and cancers (Shenderov 2015a; Althaus et al. 2013; Nicolson et al. 2016; Oleskin and Shenderov 2016; Savidge 2015). The first research into the problem have shown that ill and healthy people prescribed milk or water enriched with higher concentrations of molecular hydrogen experienced marked positive effect at various pathological conditions and increased exercise (Nicolson et al. 2016).

Hybrid Multicomponent Microbial Molecules as a Base for Metabiotics

Multi-year research into the metabolome of commensal and symbiotic microorganisms have shown that low molecular weight compounds of microbial origin can interact with different receptors of mammal microorganisms and eukaryotic cells including humans, at the same time modifying several different functions and biochemical reactions in individual bacteria, in their symbiotic associations, as well as in mammal organisms. Such polymodal effectors with respect to different human bacteria, cells, tissues and organs can be exemplified by microbial SCFA (butyrate, acetate, propionic acid etc.). In low concentrations, representatives of this group of microbial lipids can simultaneously suppress and optimize the growth of various microorganisms in the digestive tract, modify the differentiation, proliferation or apoptosis of intestinal epithelial cells, regulate the barrier function of intestinal epithelium, take part in carbohydrate, lipid, energy, water-salt and electrolyte metabolism of both individual cells and the whole organism; some fatty acids or their complex interfere with redox balance, with the synthesis of LuxS proteins and quorum-sensing regulation, gene expression and other cell processes both in *in vitro* experiments on tissue cultures and on conventional and gnotobiotic animal models. In physiological concentrations, butyrate synthesized by symbiotic (probiotic) bacteria takes part in energy homeostasis of intestinal epithelial cells and neurons in the brain, changes mood and behavior. Acetate is the key substrate for acetyl-CoA synthesis, it is involved in epigenetic processes of histone acetylation, it influences long-term memory, cell consolidation and neuroprotection/regeneration in brain tissues. Propionic acid of microbial origin can regulate human social behavior, alterate cell phospholipid composition, neuropeptide levels (serotonin, glutamate, dopamine) in cerebral vessels.

Many SCFAs interfere with the metabolism of mono-carbohydrate compounds responsible for methylation of DNA and nerve cell histones (Shenderov 2013b; Shenderov 2017; Engevik and Versalovic 2017; Mischke and Plosch 2016; Shenderov 2016b). Now the information is available that helps explain how individual microbial low molecular weight bioactive compounds or their groups can simultaneously cause different effects in prokaryotic and eukaryotic cells. It is known that the structure of bacteria and eukaryotic cells is essentially different. In mammals, cell organelles (the nucleus, mitochondria, ribosomes, exosomes etc.) are separated from each other by own membrane formations; the integration, coordination and regulation of these organelles work in the cell cytoplasm is realized through special information regulatory network structures using a multitude of signaling mediators (ion channel proteins, various kinases, transport proteins, organic acids etc.). By contrast, bacteria and archaea have no differentiated cell organelles: their role is functionally played by special formations (so-called hyperstructures) which feature temporary, complex assemblies of different molecules and macromolecules with different specific functions. Such hyperstructures in bacteria and archaea are defined as functionally dependent structures (FDS); they only exist in a bacterial cell until functionally needed. FDS may include sets of different low molecular structures, co-factors, proteins, peptides, organic acids, signaling and other molecules which interact with one another, often getting simultaneously involved in several particular functions and metabolic reactions. During the bacterial cell development cycle, in case of inducer liquidation, FDS fall into separate molecules, which subsequently can get involved into the same or new microbial hyperstructures responsible for the corresponding function and/or metabolic reaction and generation of the final metabolic product.

A multitude of FDS are simultaneously formed, function or disintegrate in the cells of bacteria and archaea. By now, functionally dependent structures have been identified in bacteria and archaea, which are responsible for the detection and transportation of food substrates and co-factors, synthesis and functioning of cell membranes, formation of cytoskeleton, synthesis of various enzymes and signaling molecules, transcription, translation and degradation of DNA and RNA, for chromosomal segregation, cell division, metabolism, polyphosphate synthesis, virulence, lipopolysaccharide synthesis, chemotaxis and other functions and biochemical reactions of microbial cells (Ghisian et al. 2018; Legent and Norris 2009; Norris et al. 2016). Through the example of *E. coli* and bacilli it was revealed that their membrane FDS may simultaneously include a multitude of various molecules (for instance, over 1000 membrane proteins alone have been identified) involved and coordinating the processes related to bacterial cell growth and division. For example, when glucose, which is a source of energy for bacterial cells (*E. coli*), is added into the culture medium, transcription occurs in many genes involved in import and synthesis of polyamines, inorganic phosphate and ions of magnesium, which ensure the functions of membrane hyperstructure of these bacteria. When bacilli enter stationary phase, the FDS of bacterial membranes show increased levels of micronutrients involved in the utilization of glycerol, ribose, lactate, nucleoside, succinate, fumarate and zinc; at the same time, quantitative decrease is observed in proteins transporting malate, citrate, hydroxamate of iron and hydroxymethyl-thiamine/thiamine. Upon completion of these processes, at the time of bacterial cell division, the membrane breaks down and a new membrane FDS appears which may differ by its component molecules, considering the particular factors of the medium surrounding the bacterial cell.

The affinity of molecules and ions inside and between hyperstructures ensures the work and coordination of all microbial FDS (Matsumoto et al. 2015; Thellier et al. 2006). Prokaryotic organisms can form both the simplest hyperstructures (sometimes consisting of two or three elements), and complex FDS assemblies, capable of generating particular metabolic signals (Thellier et al. 2006). Separate molecules, their complexes and even whole microbial FDS can be passed among various cells of one species and also among the representatives of other species (for example, among *B. subtilis, E. coli, S. aureus*). Intercellular molecular exchange occurs via cell pores (*nanotubes*), which contributes to the formation of new FDS assemblies in different bacterial cells and even in whole consortia of symbiotic bacteria in any given ecological niche (Buffie and Pamer 2013). Bacterial FDS (sometimes of a gigantic size) and their individual molecular components and complexes formed by combining dozens to thousands of simple molecules, may subsequently be transmitted and get involved in the work of metabolic and signaling processes in bacterial and archaeal FDS, and also, supposedly, in the formation and functioning of organelles in mammal and plant cells (Table 9) (Shenderov 2017; Browne et al. 2016; Buffie and Pamer 2013).

For targeted delivery of complex hybolites into human cells, the use of cell exosomes or cationic lipid complexes (*lipoplexes*) is recommended.

Table 9 Examples of complex hybrid metabolites (hybolites) with targeted effects on the human organism (Shenderov 2017; Shenderov 2018; Legent and Norris 2009; Norris et al. 2016)

Item No.	Hybolites
1	Artificially constructed hibolite on the base of microbial cardiolipin (phospholipid of bacterial membrane) and bacterial transport protein (flotillin-1) that regulates normalization of functioning of human cell membrane channels
2	Complex hybolite on the base of microbial D-isomers of leucine, methionine, tyrosine and tryptophan isolated from *B. subtilis* which inhibits staphylococci and *P. aeruginosa* biofilm formation
3	Complex hybolite on the base of serine, threonine, sarcosine and phosphocholine of bacterial origin having a potential anticancer effect

Other Contemporary Examples of the Techniques Needed to Create, Maintain and Correct Human Microbial Ecology

In the future, microecological techniques of symbiotic microbiota correction will be developing and improving. The perspectives are opening for the construction of new metabolomic dietary supplements and functional foods intended to support the genome and microbiome stability (metabolomic biotics), low molecular weight modulators of information exchange between bacteria and the host cells (informobiotics), complex modulators of development mechanisms for colorectal and skin neoplasms and their metastatic spreading into different human tissues and organs. Work will continue to create metabiotics on the base of microbial low molecular weight compounds in combination with different low molecular weight plant components (for instance, bioflavonoids). New generations of biotherapeutics with pharmaceutic activity are being created on the base of live commensal anaerobe bacteria of human origin. Such metabiotics will serve as fortifiers in functional foods, as a part of the formulation of metaprobiotics, meta-autoprobiotics, metaprebiotics, meta-synbiotics, synthetic metabiotics, and as pharmaceuticals and cosmetic products. In some countries, national cryogenic banks of human microbiocenoses will be established for long-term preservation of individual microbiota (Vakhitov et al. 2005; Volkov et al. 2007; Shenderov 2015b; Shenderov 2016a; Shenderov et al. 2018a).

Among new approaches and techniques of symbiotic microbiota correction some new technologies should be mentioned that are not directly connected with the concept of metabiotics. Of these, for over twenty years the attention has been focused on the so-called autoprobiotics, whose development requires the arrangement of cryogenic banks of individual symbiotic microbiocenoses of the digestive tract in healthy people in different countries and regions of our planet. The importance of such banks is determined by a great ethnic diversity of the human population, dietic peculiarities and the composition of symbiotic microbiota in its individual representatives, as well as by ecological and geographical peculiarities in their places of residence. Symbiotic associations of microorganisms preserved for long periods at low temperatures (from −80 to −196°C) can serve as a base for autoprobiotics for individual use or for microecological engineering (Shenderov 2003; Shenderov 2011b; Shenderov et al. 1996; Shenderov et al. 2018b; Suvorov et al. 2018). Collections of people's natural individual microbiocenoses allows for long-term preservation of biodiversity of microbiota in the human population currently living on our planet, and for manufacturing autoprobiotics by using autostrains of indigenous microbiota designed for people from whom these microbiocenoses had been isolated. If required, the preserved microbiocenoses can become a base for preparing starter cultures for manufacturing special-purpose probiotics/metabiotics for correction of changes in a particular person's metabolome that serve as potential disease risk markers or indicators of the already developed pathological condition.

A person's epigenetic program is formed in the first thousand days of his/her life; the programming process is influenced by numerous factors and agents — dietary and microbial compounds, in the first place. In this respect, it is doubtlessly rational to artificially form a mother's and a child's symbiotic microbiota (microecological engineering) (Shenderov 2007; Shenderov et al. 1996). Cryogenically preserved microbiocenoses can be used for microecological engineering – primarily, for pregnant and breastfeeding women, for neonates and infants in the first two years after birth – for controlled shaping of an individual's microbial ecology. In cases when using autoprobiotics in this category of people is impossible or impracticable, they will be substituted by individual special-purpose metabiotics (Shenderov 2003; Shenderov 2007; Shenderov 2011b; Shenderov and Midtvedt 2014). In the USA and in some Western European countries, in Japan and in Australia, chronic inflammatory diseases of the digestive tract (primarily, in the treatment of people with ulcerative colitis caused by *Clostridium difficile*) are treated by using the technology of colonic microbiota transplantation right at the patient's bedside. The technology consists in taking healthy people's fecal material and its peroral (via a special plastic tube) or intrarectal (via an enema) administration to patients with severe clinical manifestations of colitis. Usually, the number of such procedures does not exceed 10 within a period of two or three weeks. Though over 10,000 fecal transplants have been performed worldwide, they are increasingly criticized due to frequent complications following such procedures (Kazerouni et al. 2015). Negative consequences of the proposed fecal transplant technology are quite predictable, as in the vast majority of cases the procedure of transferring diluted feces fails to take into account strictly individual nature of each person's gut symbiotic microbiota (Shenderov 2014b). Another reason for failure is the fact that it is quite impossible today to identify a priori a "healthy" donor. Cryogenic banks created for preserving people's microbiocenoses in early childhood, as well as complex autoprobiotics, are the technologies that could solve all the problems emerging today in case of traditional fecal matter transplantation (Shenderov et al. 2018a, Shenderov et al. 2018b).

To date, over 500 genes have been identified in microorganisms including symbiotic ones that determine the synthesis of toxins and other pathogenicity factors; also, around 200 genes are known which are responsible for resistance to different antimicrobial compounds including antibiotics and other pharmaceuticals with antimicrobial effects (antihistamines, antidepressants etc.). In this connection, the use of live probiotics may be an important condition responsible for spreading transmissible genes of virulence and antibiotic resistance among the human population. Some scientific communities are discussing the opportunity of probiotic creation by using synthetic biology techniques. This molecular technology comprises in detailed examination of strains of symbiotic microbiota representatives for the presence in potential starter cultures of the following genes: of antibiotic resistance, pathogenicity or those responsible for somatic cell malignization, as well as genes responsible for compatibility with indigenous microbiota and immunological tolerance. After detailed examination of the genomes of potential starter cultures by using the required modern molecular genetic technologies, it is suggested to construct synthetic probiotic strains devoid of genes responsible for the synthesis of compounds with potentially harmful health and environmental effects. Simultaneously, these cultures may take genes that will be introduced to determine the synthesis of bioactive compounds potentially beneficial for human health. Actually, what this means is the development of new microbiological cultures including those that have never existed in nature (Shenderov 2014b). The search for new locally active potential antimicrobial compounds (phagobiotics, bacteriocins, other antimicrobial peptides, new versions of antibiotics etc.) should be viewed as one more line for correction of unbalanced microecology of the digestive tract. Its implementation can also be based on microorganism strains kept in cryogenic banks of healthy people's individual microbiocenoses (Shenderov 2014b; Shenderov et al. 2014; Shenderov et al. 2018a; Shenderov et al. 2018b; Sanshez et al. 2015; Venema and Carmo 2015).

Conclusion

As regards cell quantitative content, a human being is essentially a more prokaryotic than eukaryotic organism. Great success has been achieved in the characterization of microorganisms inhabiting the human being, all their genetic elements have been explored in detail by using leading-edge molecular sequencing techniques. The state of "host-its microbiota" system is one of the leading factors that determine the growth, development and health of a human being; the importance of the microbiome in the pathogenesis, phenotype, prognosis of many diseases has been established. Symbiotic (indigenous) microorganisms inhabiting his/her skin and mucosa are a source of various low molecular weight compounds involved in environmental perception, in bacterial information intra- and intercellular exchange both among themselves and with the host cells, in the regulation of physiological functions, biochemical and behavioral responses. Different strains form their own strain-specific range of similar substrates, co-factors, enzymes and regulatory molecules which bear chemical and functional similarity with those produced by the host's eukaryotic cells and numerous micronutrients contained in food products. It means that many compounds of microbial origin should be viewed as universal low molecular weight substrates and co-factors of epigenetic, metabolic, immune, neurohormonal reactions, diffuse regulatory endocrine system (APUD system) and the participants of intra- and interpopulation communication between bacteria and the host eukaryotic cells. Bioactive molecules of microbial origin are capable of interaction with the host cells through binding to the receptors of membranes, mitochondria, ribosomes. These molecules can also be accumulated in intracellular exosomes and regulate intercellular and intracellular channels, influencing the structure, synthesis, transfer and function of regulatory and signaling molecules (Aguilar-Toaláa et al. 2018; Bik et al. 2018; Engevik and Versalovic 2017; Gilbert et al. 2016; Lebeer et al. 2018; Maguire and Maguire 2019; Shenderov 2011a; Shenderov 2011b; Singh et al. 2018).

Unfortunately, in the last fifty years, numerous biotic and abiotic exogenous and endogenous stress factors affecting the organism have disturbed the balance between human symbiotic, eukaryotic and prokaryotic cells, have induced marked

B. A. Shenderov et al., *METABIOTICS*,
https://doi.org/10.1007/978-3-030-34167-1_17

modifications in the composition and functions of its microbial symbiotic component, which has led to the increased risk of many chronic metabolic disorders. Initially, these damaging agents usually cause transitory negative changes in the functioning of the "host-its microbiota" ecosystem, which are then followed by metagenome, metaepigenome and information disorders. If the exposure is long-lasting enough, so that it exceeds the system's compensatory capabilities, various pathological syndromes and conditions develop. Over the last 50 to 100 years, anthropogenic negative impact on the environment, as well as the man-of-today's unbalanced diet came into collision with his ability to adapt. That is why these two particular factors are now seen as two main causes of the progressively imbalanced microecological profile in a good number of representatives of the human population.

Inadequate stress loads observed for several generations have induced deregulation of those dietary and gut microbiome functions that formerly ensured human dietary, microecological, metabolic, epigenetic, neurohormonal and immune homeostasis, and have increased the risk of negative epigenome modifications related to premature ageing and degenerative metabolic diseases (Shenderov 2008; Shenderov 2011a; Shenderov 2011b). In order to prevent and restore these disorders, various microecological techniques and technologies, as well as particular probiotic therapeutics and food products, have been proposed and practically implemented in the last decades. Unfortunately, long experience of their use has shown that traditional probiotics have a wide range of drawbacks related, in the first place, to their safety in case of long-term use.

Knowledge accumulated over many years about the nature of man as a peculiar kind of superorganism consisting of a multitude of symbiotic eukaryotic and prokaryotic cells and viruses, about the role and molecular language of symbiotic microorganisms in maintaining human health, as well as cellular and molecular mechanisms of contemporary metabolic diseases, allowed to define a new biotechnical trend of "METABIOTICS." It consists of using metabolomic potential of symbiotic (probiotic) microorganisms for commercial production of unique therapeutics, dietary supplements and functional foods on the base of low microbial weight substances of a known chemical structure having diverse biological and/or pharmacological systemic or selective effects, as well as good human safety profile. These microbial bioactive compounds can be used as biotechnological raw materials for commercial production of microecological products free from viable microorganisms, different in composition and in methods of administration, for prevention and treatment of chronic conditions in humans, as well as for replenishment of various nutritional deficiencies in sportsmen and old people. Presently, the world's markets of therapeutics, functional foods, dietary supplements can offer more than twenty different metabiotics on the base of both microbial cell substances and various low molecular metabolite, regulatory and signaling molecules (Table 15).

Various *Bacillus* species are among the most powerful producers of bioactive secondary metabolites (up to 800 compounds), which allows to consider the strains of these species as "the most perfect multifunctional probiotic bacteria." The said potential has been realized by Russian developers in Bactistatin produced by using the *Bacillus subtillis* strain No.3. Bactistatin is a germfree filtrate of the *B. subtilis* cultural fluid containing a complex of microbial low molecular weight compounds immobilized on zeolite in admixture with soy flour hydrolysate. This product has been proven in human clinical trials, normalizing microbial ecology of the digestive tract, improving digestion, showing immune modulating and detoxicating effects, normalizing motor and evacuation functions of the intestine. This product improvement is attempted along the lines of creating combined therapeutics containing added active target-focused ingredients.

One example of metabiotics on the base of cell components of probiotic lactobacilli is *Pylopass*, made to control the *Helicobacter pylori* levels in the stomach and to prevent the related disorders. The examination of over 700 lactobacilli conducted in Germany allowed to isolate the *Lactobacillus reuteri* DSM 17648 strain which is capable of binding to the *Helicobacter pylori* surface and cause the formation of bacterial aggregates, which neutralized the activity of this pathogen and allowed for eliminating it from the digestive tract of an infected individual (Hunt et al. 2011). Moreover, no viable Pylopass cells are needed for the specific anti-*H. pylori* action — lyophilized or spray dried cells are sufficient. The resulting complex of *Lactobacillus–Helicobacter* is naturally evacuated from the stomach. It contributes to considerable reduction in *Helicobacter* levels and offers an essentially new approach to controlling bacterial load in the stomach after or in parallel with eradication therapy and as a prophylactic supplement. To date, it is a unique product with a proven mechanism of action against *Helicobacter pylori*. The absence of live lactobacilli considerably reduces the risk of potential side effects typical of probiotics containing live strains; besides, it ensures good product stability for not less than three years.

In the future, metabiotics will serve as a base for developing semi-synthetic or synthetic metabiotic analogues. Antibiotics industry developed in much the same way in its time (Shenderov et al. 2010; Shenderov 2011a). When practically implementing the concept of "METABIOTICS," it is critical to conduct detailed studies of the metabolome of all known and potential probiotic strains in order to provide an opportunity of using them in the future as starter cultures for commercial production of microbial substances for different pharmacological or dietic applications. Dozens and hundreds of microbial bioactive compounds – both known and new – may be used as metabiotics in the years to come (Shenderov 2009; Shenderov 2017; Shenderov et al. 2010; Cani 2018; Engevik and Versalovic 2017; Lebeer et al. 2018; Maguire and Maguire 2019; Shenderov 2011a). As compared to traditional probiotics on the base of live organisms, metabiotics have quite a few advantages: they are characterized by known chemical structure, clear application targets; they are much easier to dose, and to control as regards safety and pharmacological and biological activity. They are more easily absorbed, metabolized and distributed to tissues and organs, and more quickly removed from the organism. Finally, they have longer storage periods (Aguilar-Toaláa et al. 2018; Osipov and Verkhovtseva 2011). Also, bioactive compounds produced by the representatives of human symbiotic microbiota compare favorably to chemically alike substances of dietary or other origin.

For millions of years of evolution, human superorganism had been selecting prokaryotic and eukaryotic microorganisms from the environment, which were optimal for its life and development from the functional and biological perspective. That is why metabiotics constructed on the base of bio- and pharmacologically active low molecular weight compounds of human symbiotic microorganisms will undoubtedly be most beneficial, effective and safe for health. Exact determination of the quantitative content of microbial compounds with specific functional, metabolic or signaling activity in human tissues and biological fluids will also help trace causal relationship between the particular types of microecological imbalance and pathological syndromes and diseases (Dorrestein et al. 2014; Gilbert et al. 2016; Shenderov 2011a; Singh et al. 2018; Sonnenburg and Backhed 2016). The next generation of metabiotics (semi-synthetic, synthetic) will further promote the concept of probiotics, improving efficiency, specificity and safety of classic probiotic effects; they will also reduce the risks associated with currently used microecological techniques of prevention and treatment of diseases related to the imbalance of mammal symbiotic microbiota.

Speaking about metabiotics, it should be remembered that this applied scientific biotechnical trend is now at the earliest stage of its development. To date, the range of gut microorganisms capable of forming colonies on artificial culture media is barely over 1500 species (Browne et al. 2016; Rajilic-Stojanovic and de Vos 2014). Many hundreds and thousands of species of other germs colonizing human digestive tract are yet to be isolated. Even such symbiotic microorganisms as lactobacilli have not yet been fully characterized taxonomically (over 230 *Lactobacillus* species have been identified; around 10 new species are being identified annually). Many researchers, technologists and clinicians have insufficient knowledge of the quantity and spectrum of newly found microbial low molecular weight compounds, their detailed physical and chemical characteristics, targets and molecular mechanisms of action, factors and conditions that influence their positive activity and side effects on a microorganism, especially when using potential metabiotics in inadequate concentrations. Probably, when discussing metabiotics – these new forms of therapeutics – functional foods or dietary supplements, it should be remembered that it is necessary to create new adequate model systems highly sensitive to low concentrations of these microbial compounds and their complexes. We suppose that for these purposes could be used available banks of different eukaryotic cell lines obtained from conventional and germfree animals, and also new model subcellular structures may be created (membranes, mitochondria, ribosomes, exosomes, ion channels) including those that will be created through synthetic biology techniques. This will enable to determine, on molecular level, the effects of low metabiotic concentrations on energy, protein, lipid and carbohydrate metabolism of cells, on the operation of their membranes, mitochondria, ion channels etc. The results obtained by using these models will be even more informative if the newly created metabiotics contain low molecular weight compounds carrying labeled isotopes in their structure.

When constructing metabiotics, the primary requirement is for up-to-date analytical base that allows for expression and high-precision evaluation of raw products for subsequent extraction of microbial ingredients, final metabolic products, the content of the corresponding functional ingredients, their physical and chemical characteristics that influence bio-accessibility and final specific effects (Table 1).

It is important to know the exact targets and functional ingredients of dietary and endogenous (cellular and microbial) origin and the conditions that can bring about the most pronounced favorable (or negative) effects. The use of various methods from OMIC-technologies, germfree and gnotobiotic animals of different levels of organization, as well as new and synthetic models of cell cultures and organelles isolated from them will allow specialists that study and develop intended-use metabiotics to achieve even greater success in realizing this trend in practical medicine and nutritiology. Russia is one of the first countries having taken stock of biotechnological prospects and importance of creation of various-purpose metabiotics that can prevent, restore and regulate physiological functions, biochemical and behavioral responses, signaling intra- and intercellular communication, epigenetic regulation of gene expression and post-translational modification of their final products. Unfortunately, metabiotics range and production volumes in our country are clearly out of keeping with the importance of these microecological means of health support and preservation in humans, animals and plants. The difficulty of bulk development and application of various-purpose metabiotics in the Russian Federation is caused by a number of reasons:

– the diversity of biological and pharmacological effects of many known low molecular weight compounds of microbial origin;

Table 1 Factors influencing functional properties and bio-accessibility of physiologically active ingredients of microbial origin included in the composition of metabiotics

Item No.	Factors
1	Physical and chemical characteristics of an ingredient: – Molecular structure (L- or D-forms, α-, β-, γ- or other forms of a molecule; valency and isotopic state of chemical elements included in the structure of the functional ingredient etc.). – Solubility, dispersity, bonds with ligands of macro- and micronutrients. – Oxidation state. – Interaction with other components (amplifiers and inhibitors of ingredient effect). – Competition for specific transport proteins or absorption sites.
2	Physiological conditions related to the host organism: – Gastric acidity. – pH and redox potential in the gastric lumen. – Metabolic activity of intestinal juices. – Bowel motility. – Digestive and physiological status (age, sex, pregnancy, lactation etc.). – Different stress factors. – Hereditary defects.
3	The state of symbiotic microbiota of the digestive tract.
4	The technologies of obtaining functionally active microbial connections.
5	Other factors.

- the absence of economically sound techniques of industrial isolation and assessment of physical and chemical structure of potential metabiotics, as well as the interaction of their biologically and pharmacologically active microbial compounds with other low molecular weight compounds of dietary or endogenous origin;
- insufficient use of OMIC-technologies and gnotobiotic models in our country for the assessment of safety and efficiency of this new class of dietary supplements and pharmaceuticals of microbial origin;
- lack of biotechnological factories and skilled personnel educated in this sphere;
- not enough museums of microorganisms and microbiocenoses of human origin and microbial materials preserved therein.

Some of the above statements may be illustrated by the following examples. It is known that in course of carbohydrates, poly- and oligosaccharides, proteins and lipids fermentation, symbiotic microorganisms ever present in the digestive tract form a considerable amount of SCFAs; participating in energy metabolism, these fatty acids also display a multitude of other biological effects. SCFAs suppress the growth of pathogenic and opportunistic bacteria, control gut barrier function, regulate cell differentiation, proliferation, apoptosis and mucous layer renewal, optimize the growth and development of probiotic and symbiotic bacteria, take part in carbohydrate and lipid metabolism, in water-salt exchange, in production and regulation of hormone and neurotransmitter activity, cause immune effects, eliminate damaged cells, prevent neoplastic transformation of normal cells, take part in LuxS protein synthesis, in epigenome regulation of gene expression, quorum-sensing regulation (Golovenko et al. 2011; Carding et al. 2015). Naturally, using them as metabiotics will require that producers and doctors have full knowledge on particular targets these microbial SCFAs are planned to be used against, concentrations, forms and lengths of their administration etc.

Nowadays, there are over 600 different collections of microorganisms. But only about 20 of them have the status of internationally recognized microorganism collections, since only they are capable of performing the functions required from this type of organizations. Their maintenance demands considerable investments from the countries' governments. To give an example, one could mention German Collection of Microorganisms (DSMZ). Prokaryote, fungi and yeast section in DSMZ supports 26 thousand cultures with 6 scientists and 18 employees on service. In 2011, the cost of keeping this collection amounted to around EUR three million; it distributes over 20,000 strains annually (40% domestically and 60% abroad). To maintain the collection and conduct scientific research, about 1500 different culture media are used. The first low-temperature bank in the Russian Federation for keeping pure microorganism cultures was established at the Institute of Microorganism Biochemistry and Physiology of Russian Academy of Sciences on the basis of All-Russian Collection of Microorganisms. Unfortunately, though this Russian collection has been registered at the World Registry of Microorganism Collections, it is not included among those complying with international requirements (Shenderov et al. 2014).

Cryogenic banks of natural microbiocenoses of humans, animals, plants, surface water bodies and soils deserve special mentioning. Cryogenic banks help preserve intact (in viable state) symbiotic microbiocenoses of soils, water bodies, any cellular eukaryotic organisms having emerged in course of the evolution for extended periods (practically for eternity) at low temperatures. The microorganisms kept in cryobanks can subsequently be used for making personified foods, pharmaceuticals, for maintaining and restoration of microbiocenoses in people of hazardous occupations, in the production of targeted microbial fertilizers designed for particular plant species, in the preservation and restoration of rare or extinct animals and plants, in the creation of specific pollutant destructors, in the remediation of soils and water bodies, in the implementation of other cutting-edge biotechnological trends using the genetic and biochemical potentials of the representatives of symbiotic microbiota of different organisms. World's first cryobank of natural microbiocenoses was established in 1995–1996 on the basis of genetic resource cryopreservation laboratory at the Institute of Cell Biophysics, Russian Academy of Sciences (Pushchino-on-Oka town, Russia). It was intended for long-term preservation of human colon microbiocenoses. In concert with the researchers from Science and Innovation Center of "Russian Yoghurt", JSC, and later with the assistance from G.N. Gabrychevsky Moscow Research Institute of Epidemiology and Microbiology, the laboratory personnel had chosen optimal cryopreservation agents, worked out viability assessment techniques for the preserved microorganism complexes, developed the cryobank structure, selected the necessary equipment, prepared a complete feasibility report for establishing a cryoreservoir for 20,000 intestinal content samples. Biomaterial obtained from the children of cryopreservation laboratory personnel at the Institute of Cell Biophysics, Russian Academy of Sciences, was placed for long-term storage. Multi-year research into the cryopreservation biomaterial had shown that the invented technology for storage of natural human gut microbiocenoses is fully compliant with the requirements for live organism cryopreservation (Shenderov et al. 1996; Shenderov et al. 2014; Shenderov et al. 2018a; Shenderov et al. 2018b; Shenderov et al. 2018c).

Presently, a cryobank of natural human microbiocenoses containing dozens of thousands of biomaterial samples was established and is functioning in France. The groups of people whose symbiotic microbiocenoses are supposed to be sent for long-term storage in cryobanks, should include, in the first place, all healthy children aged 2 to 6, as well as young women (primarily, first-time mothers). Such cryobanks should be created for people working for a long time under extreme conditions (law enforcement personnel, pilots, submariners, astronauts, journalists, public transport drivers, businessmen etc.), as well as for people living or working in hazardous environments (chemical factory and nuclear power plant employees, Arctic expedition participants etc.); practically all people willing to preserve their natural microbiota for long periods of time and use it, if necessary, for microbial autotransplantation would be able to send their natural microbiocenoses to such cryobanks (Shenderov et al. 2018a; Shenderov et al. 2018b; Suvorov et al. 2018).

Considering the ever-increasing importance of symbiotic microbiota for preserving our planet's biodiversity, for maintaining health and reducing the risks of so-called "lifestyle diseases," one can easily understand why so much attention is being paid worldwide to the construction of effective and economically sound microecological strategies. Accumulated knowledge regarding the molecular language of symbiotic microorganisms allows for constructing more effective and targeted microecological products (therapeutics, dietary supplements, functional and personal foods) for prevention and treatment of many metabolic diseases associated with microecological disorders, changes in the epigenetic remodeling of chromatin, DNA and proteins. In the future, microecological techniques of symbiotic microbiota correction will be developing and improving (provision is made for the creation of metabiotics on the base of microbial low molecular weight compounds individually and in combination with various plant bioflavonoids; development of national cryogenic banks of human microbiocenoses for long-term preservation of individual microbiota, for constructing personal-use metabiotics and microecological engineering; construction of probiotics and metabiotics by using the techniques of synthetic biology; creation and implementation of new types of local-action antimicrobial compounds.) Subsequent generations of simple, complex, semi-synthetic and synthetic metabiotics (by analogy with antibiotics) will be analogues or improved copies of natural low molecular weight compounds of probiotic and commensal human microorganisms. It is beyond argument that before introducing the specified groups of metabiotics into preventive and clinical medicine they should be subjected to thorough pre-clinical trials, which must take into account the peculiarities of their physical and chemical structure and mechanism of action on any given targets of the host's tissues and organs.

We suppose that in the years to come the following techniques for maintaining and restoring human microbial ecology will be gaining momentum:

- General-purpose, specific and personified metabiotics on the base of individual and hybrid microbial low molecular weight compounds – individually and in combination with different plant nutrients, regulatory and signaling molecules.
- Cryogenic banks of human microbiocenoses for long-term preservation of individual microbiota, for the construction of antibiotics for personal use and microecological engineering.
- Probiotics made by using the techniques of synthetic biology.
- New types of locally active antimicrobial compounds (phagobiotics, bacteriocins, other antimicrobial peptides, new versions of antibiotics etc.).

It is about time an international interdisciplinary program is created: "METABIOTICS — NEW NUTRITIVE AND MICROECOLOGICAL STRATEGY OF ACTIVE LONGEVITY AND PREVENTION OF CHRONIC METABOLIC DISEASES," whose implementation will allow for a dramatically decreased risk and progression of the main "lifestyle diseases." It could become a useful, fruitful and consolidating program that would actively attract a wide range of Russian scientists, biotechnologists and clinicians in diverse areas of expertise.

The transformation of probiotic strains into metabiotic therapeutics is an effective strategy intended to ensure high-quality active ingredients availability in a stable pharmaceutical formulation. Understanding the mechanism of action of probiotic therapeutics responds to the needs of society and science, lays the foundation of the development of new, safe and highly effective bioactive therapeutics. To better understand human interaction with low molecular weight bioactive compounds of microbial origin, experimental and clinical studies should be increasingly often based on the leading-edge "multiOMIC technologies" (genomics and its varieties — pathogenomics, probiogenomics; transcriptomics and its variety — RNAomics; proteomics and its variety — interactomics; metabolomics and its variety — fluxomics; epigenomics; phenomics). Simultaneous experimental use of different germfree animals and gnotobiotes including those inoculated with symbiotic microorganisms taken from people of different ages, sex and races will help obtain the most objective data on metabiotics application targets (Shenderov 2012; Shenderov 2014a; Shenderov 2014b; Dorrestein et al. 2014; Karczewski and Snyder 2018; Martin et al. 2008).

When evaluating the interaction of humans with their symbiotic microbiota in ill and in healthy people, it should always be remembered that every person is individual both on genome and microbiome level. Preventive medicine for each particular person should be based on the knowledge about the peculiar features of his/her genome and microbiome, on the understanding of molecular mechanisms of communication between them and the role of environmental factors in developing, maintaining and epigenetic modification of the host's total metagenome functions. Sudden and long-lasting defects of energetic metabolic processes in mitochondria and bacterial membranes are often related to the deficiency or excessive ingestion of substrates, co-factors or enzymes and disturbance of functioning of the microbiota–mitochondria–epigenetics axis. It is accompanied by insufficient ATP generation, increased content of high-reactive free radicals of oxygen, nitrogen and products of peroxidation in cells, imbalance in antioxidant support network and compounds involved in processes of gene expression regulation and in post-translational modification of final products of these genes. As a consequence, chronic inflammation is induced, telomeres become shorter, processes of proteostasis are lost, cell ageing occurs, stem cell pool diminishes, intercellular communication is disrupted – all this results in quick ageing and risks of chronic pathological conditions (neurodegenerative, metabolic, autoimmune, behavioral, mental, musculoskeletal, inflammatory diseases, chronic infections, cancer) (Paul et al. 2015; Singh et al. 2018).

There are no food products, pharmaceuticals, physiological functional ingredients, which would have absolutely identical effect on everyone. Healthy and tailored diet that takes into account a person's individual needs helps restore the expression of the polymorphic genes with adverse health effects, and, on the contrary, to optimize epigenomic control and realization of those allelic genes that ensure better adaptation of the human "superorganism" and harmonize the work of its symbiotic components in normal conditions, under stress and extreme conditions. Understanding of this fact helps to further increase specificity, efficiency and safety of microecological techniques aimed at prevention and treatment of diseases related to mammal symbiotic microbiota imbalance (Shenderov 2017; Sharma and Shukla 2016; Singh et al. 2018).

To date, we are only at the beginning of the new era of molecular personal bio-therapy and nutrition. Nevertheless, it is beyond doubt that soon it will be possible to intentionally and successfully manipulate both humans and their microbiota as a result of interfering with their informational interaction, stability and epigenome gene expression regulation by using different types of low molecular weight compounds of eukaryotic, prokaryotic and dietary origin (Shenderov 2011a). Even present-day knowledge on the structure and functions of human symbiotic microbiota in all periods of life and development calls for revolutionary changes in the training of future medical professionals with shifting the focus from this paradigm – “a human being represents a community of eukaryotic cells” to another paradigm – “a human being represents a superorganism, a community of eukaryotic, prokaryotic cells and viruses”; given that, professional training programs should highlight that the state of symbiotic microbiota is really the leading factor in human health.

Bibliography

Aguilar-Toaláa JE, Garcia-Varela R, Garcia HS, Mata-Harod V, González-Córdovaa AF et al. Postbiotics: An evolving term within the functional foods field. Trends in Food Science & Technology. 2018;75:105–114. doi:https://doi.org/10.1016/j.tifs.2018.03.009.

Almada CN, Almada CN, Martinez RCR, Sant'Ana AS. Paraprobiotics: Evidences on their ability to modify biological responses, inactivation methods and perspectives on their application in foods. Trends in Food Science & Technology. 2016;58:96–114. doi:https://doi.org/10.1016/j.tifs.2016.09.011.

Althaus M, Behnsen J, Deriu E, Sassone-Corci M, Raffatellu M. Probiotics: properties, examples, and specific applications. Cold Spring Perspect Med. 2013;3:a0110074. doi:https://doi.org/10.1101/cshperspect.a010074.

Ardatskaya MD Probiotics, prebiotics and metabiotics in the correction of microecological bowel disorders. Medical Council. 2015;3:94–99 (in Russian).

Ardatskaya MD, Belmer SV, Dobritsa VP, Zakharenko SM, Lazebnik LB, Minushkin ON, Oreshko LS, Sitkin SI, Tkachenko EI, Suvorov AN, Khavkin AI, Shenderov BA. Colon dysbacteriosis (dysbiosis): modern state of the problem, comprehensive diagnosis and treatment correction. Experimental & Clinical Gastroenterology. 2015;117(5):13–50 (in Russian).

Arthur JC, Perez-Chanona E, Muhlbauer M, Tomkovich S, Uronis JM, Fan T-J et al. Intestinal inflammation targets cancer-inducing activity of the microbiota. Science. 2012;338:120–123. doi:https://doi.org/10.1126/science.1224820.

Ashraf R, Vasiljevic T, Smith SC, Donkor ON. Effect of cell-surface components and metabolites of lactic acid bacteria and probiotic organisms on cytokibe production and induction of CD25 expression in human peripheral mononuclear cells. J Dairy Sci. 2014;97(5):2542–2558. doi:https://doi.org/10.3168/jds.2013-7459.

Atkinson S, William P. Quorum sensing and social networking in the microbial world. J Royal Soc Interface. 2009;6(40):959–78. doi:https://doi.org/10.1098/rsif.2009.0203.

Aurich MK, Thiele I. Computational modeling of human metabolism and its application to systems biomedicine. Methods Mol Biol. 2016;1386:253–281. doi:https://doi.org/10.1007/978-1-4939-3283-2_12.

Beger RD, Dunn W, Schmidt MA, Gross SS, Kirwan JA et al. Metabolomics enables precision medicine: «A White Paper, Community Perspective». Metabolomics. 2016;12:149. doi:https://doi.org/10.1007/s11306-016-1094-6.

Beider C, Flambard B. Dairy for heart health. World Food Ingredients. 2003:24–26.

Beloborodova NV, Bairamov IT, Olenin AYu, Fedotcheva NI. Exometabolites of some anaerobic microorganisms of human microflora. Biomedical chemistry. 2011;57(1):95–105 (in Russian).

Beloborodova NV, Osipov GA. Small molecules originating from microbes (SMOM) and their role in microbes-host relationship. Microb Ecol Health Dis. 2000;12:12–21.

B. A. Shenderov et al., *METABIOTICS*,
https://doi.org/10.1007/978-3-030-34167-1

Beloborodova NV. Integration of human metabolism and its microbiota in critical conditions. General Reanimatology. 2012;VIII(4):42–54 (in Russian).

Belousova EF, Nikitina YuI, Mishurovskaya N, Zlatkina AR. Possibilities of microbial metabolites in the recovery of intestinal microbiota. Consilium Medicum. 2005;7:9–13 (in Russian).

Bengmark S. Ecological control of the gastrointestinal tract. The role of probiotic flora. Gut. 1998;42:2–7. doi:https://doi.org/10.1136/gut.42.1.2.

Bengmark S. Nutrition and resistance to disease. In: Heidt PJ, Nieuwenhuis P, Rusch VD, van der Waaij D, editors. Old Herborn University Seminar Monographs 14: Intestinal translocation. Herborn Litterae: Herborn-Dill; 2001. p. 117–133.

Bengmark S. Processed Foods, Dysbiosis, Systemic Inflammation, and Poor Health. Current Nutrition & Food Science. 2013;9(2):113–143. doi:https://doi.org/10.2174/1573401311309020006.

Berer K, Mues M, Koutrolos M, Rasbi ZA, Boziki M, Johner C et al. Commensal microbiota and myelin autoantigen cooperate to trigger autoimmune demyelination. Nature. 2011;479:538–541. doi:https://doi.org/10.1038/nature10554.

Bezrodny SL, Shenderov BA. Intestinal microbiota as a source of novel biomarkers of ageing. Journal of Restorative Medicine & Rehabilitation. 2016;1:21–28 (in Russian).

Bhat MI, Kapila R. Dietary metabolites derived from gut microbiota: critical modulators of epigenetic changes in mammals. Nutr Rev. 2017;75(1):374–389. doi:https://doi.org/10.1093/nutrit/nux001.

Bik EM, Ugalde JA, Cousins J, Goddard AD, Richman J, Apte ZS. Microbial biotransformations in the human distal gut. British J Pharmacology. 2018;175(24):4404–4414. doi:https://doi.org/10.1111/bph.14085.

Biocrates Life Sciences. Research products: The essence of metabolic phenotyping — accurate metabolic measurement. http://www.biocrates.com/products/research-products. Accessed 28 June 2019.

Bioray. CytoFlora®. http://www.bioray.com/cytoflora/. Accessed 28 June 2019.

Birmpa A, Sfika V, Vantarakis A. Ultraviolet light and ultrasound as non-thermal treatments for the inactivation of microorganisms in fresh ready-to-eat foods. Inter J Food Microbiol. 2013;167:96–102. doi:https://doi.org/10.1016/j.ijfoodmicro.2013.06.005.

Blander JM, Longman RS, Iliev ID, Sonnenberg GF, Artis D. Regulation of inflammation by microbiota interactions with the host. Nature Immunology. 2017;18(6):851–860. doi:https://doi.org/10.1038/ni.3780.

Blum HE. The human microbiome. Advan Med Science. 2017;62:414–420. doi:https://doi.org/10.1016/j.advms.2017.04.005.

Bomba A, Branderburova A, Ricanyova J, Strojny L, Chmelarova A et al. The role of probiotics and natural bioactive compounds in modulation of the common molecular pathways in pathogenesis of atherosclerosis and cancer. Biologia. 2012;67:1–13.

Bourdichon F, Berger B, Casaregola S, Farrokh C, Frisvad JC et al. A Safety Assessment of Microbial Food Cultures with History of Use in Fermented Dairy Products. In: Safety Demonstration of Microbial Food Cultures in Fermented Food Products. Bull Intern Dairy Federation. 2012;455:2–8.

Braune A, Blaut M. Bacterial species involved in the conversion of dietary flavonoids in the human gut. Gut microbes. 2016;7:216–234. doi:https://doi.org/10.1080/19490976.2016.1158395.

Bron PA, Grangette C, Mercenier A, de Vos WM, Kleerebezem M. Identification of *Lactobacillus plantarum* genes that are induced in the gastrointestinal tract of mice. J Bacteriol. 2004;186:5721–5729. doi:https://doi.org/10.1128/JB.186.17.5721-5729.2004.

Browne HP, Forster SC, Anonye BO, Kumar N, Neville BA et al. Culturing of "unculturable" human microbiota reveals novel taxa and extensive sporulation. Nature. 2016;533:543–546. doi:https://doi.org/10.1038/nature17645.

Buckley M, Lacey S, Doolan A, Goodbody E, Seamans K. The effect of Lactobacillus reuteri supplementation in Helicobacter pylori infection: a placebo-controlled, single-blind study. BMC Nutrition. 2018;4:48. doi:https://doi.org/10.1186/s40795-018-0257-4.

Buffie CG, Pamer EG. Microbiota-mediated colonization resistance against intestinal pathogens. Nat Rev Immunol. 2013;13(11):790–801. doi:https://doi.org/10.1038/nri3535.

Buriti FCA, Bedani R, Saad SMI. Probiotic and prebiotic dairy desserts. In: Watson RR, Preedy VR, editors. Probiotics, Prebiotics, and Synbiotics. 2016. p. 345–360. doi:https://doi.org/10.1016/B978-0-12-802189-7.00023-X.

Cain AM, Karpa KD. Clinical utility of probiotics in inflammatory bowel disease. Altern Ther Health Med. 2011;17(1):72–79.

Canani RB, Costanzo MD, Leone L, Pedata M, Meli R, Calignano A. Potential beneficial effects of butyrate in intestinal and extraintestinal diseases. World J Gastroenterol. 2011;17(12):1519–1528. doi:https://doi.org/10.3748/wjg.v17.i12.1519.

Cancer-causing gut bacteria exposed. Society for General Microbiology. Science Daily. 2008. http://www.sciencedaily.com/releases/2008/09/080921201716.htm. Accessed 28 June 2019.

Cani PD Human gut microbiome: hopes, threats and promises. Gut. 2018;67:1716–1725. doi:https://doi.org/10.1136/gutjnl-2018-316723.

Cani PD, Delzenne NM. Gut microflora as a target for energy and metabolic homeostasis. Curr Opin Clin Nutr Metab Care. 2007;10(6):729–734. doi:https://doi.org/10.1097/MCO.0b013e3282efdebb.

Cannon JP, Lee TA, Bolanos JT, Danziger LH. Pathogenic relevance of Lactobacillus: a retrospective review of over 200 cases. Eur J Clin Microbiol Infect Dis. 2005;24:31–40. doi:https://doi.org/10.1007/s10096-004-1253-y.

Carabotti M, Scirocco A, Maselli MA, Severi C. The gut-brain axis: interactions between enteric microbiota, central and enteric nervous systems. Ann Gastroenterol. 2015;28(2):203–209.

Carding S, Verbeke K, Vipond DT, Corfe BM, Owen LJ. Dysbiosis of the gut microbiota in disease. Microb Ecol Health Dis. 2015;26:26191. doi:https://doi.org/10.3402/mehd.v26.26191.

Caselli M, Vaira G, Calo G, Papini F, Holton J, Vaira D. Structural bacterial molecules as potential candidates for an evolution of the classical concept of probiotics. Adv Nutr. 2011;2:372–376. doi:https://doi.org/10.3945/an.111.000604.

Cats A, Kuipers EJ, Bosschaert MA, Pot RG, Vandenbroucke-Grauls CM, Kus-ters JG. Effect of frequent consumption of a Lactobacillus casei-containing milk drink in Helicobacter pylori-colonized subjects. Aliment Pharmacol Ther 2003;17:429–435.

Charbonneau MR, Blanton LV, Di Giulio DB, Relan DA, Lebrilla CB, Mills DA, Gordon JI. A microbial perspective of human developmental biology. Nature. 2016;535:48–55. doi:https://doi.org/10.1038/nature18845.

Chernevskaya EA, Beloborodova NV. Gut microbiome in critical illness (Review). General Reanimatology. 2018;14(5):96–119. https://doi.org/10.15360/1813-9779-2018-5-96-119.

Chervinets YuV, Chervinets VM, Shenderov BA. The modern view on the biotechnological potential of human symbiotic microbiota. Upper Volga medical journal. 2018;17(1):19–26 (in Russian).

Chilloux J, Neves AL, Boulange CL, Dumas M-E. The microbial-mammalian metabolic axis: a critical symbiotic relationship. Curr Opin Clin Nutr Metab Care. 2016;19(4):250–256. doi:https://doi.org/10.1097/MCO.0000000000000284.

Choi SW, Friso S. Epigenetics: A New Bridge between Nutrition and Health. Adv Nutr. 2010;1(1):8–16. doi:https://doi.org/10.3945/an.110.1004.

Clarke G, Stilling RM, Kennedy PJ, Stanton C, Cryan JF, Dinan TG. Minireview: Gut Microbiota: The Neglected Endocrine Organ. Molecular Endocrinology. 2014;28(8);1221–1238. doi:https://doi.org/10.1210/me.2014-1108.

Clayton TA, Baker D, Lindon JC, Everett JR, Nicholson JK. Pharmacometabonomic identification of a significant host-microbiome metabolic interaction affecting human drug metabolism. Proc Natl Acad Sci USA. 2009;106(34):14728–14733. doi:https://doi.org/10.1073/pnas.0904489106.

Corthesy B, Gaskins HR, Marcenier A. Cross-Talk between Probiotic Bacteria and the Host Immune System. J Nutr. 2007;137:781–790. doi:https://doi.org/10.1093/jn/137.3.781S.

Cruchet S, Obregon MC, Salazar G, Diaz E, Gotteland M. Effect of the ingestion of a dietary product containing Lactobacillus johnsonii La1 on Helicobacter pylori colonization in children. Nutrition. 2003 Sep;19(9):716–21.

Cryan JF, Dinan TG. More than a Gut Feeling: the Microbiota Regulates Neurodevelopment and Behavior. Neuropsychopharmacology Reviews. 2015;40:241–242. doi:https://doi.org/10.1038/npp.2014.224.

Cyr AR, Hitchler MJ, Domann FE. Regulation of SOD2 in cancer by histone modifications and CpG methylation: closing the loop between redox biology and epigenetics. Antiox Redox Signal. 2013;18(15):1946–1955. doi:https://doi.org/10.1089/ars.2012.4850.

Del-Immune V. http://www.delimmune.com/research/. Accessed 28 June 2019.

Demeure MJ. Physiology of the APUD system. Semin Surg Oncol. 1993;9(5):362–367.

Di Caro S, Tao H, Grillo A, Elia C, Gasbarrini G et al. Effects of Lactobacillus GG on gene expression pattern in small bowel mucosa. Dig Liver Dis. 2005;37:320–329. doi:https://doi.org/10.1016/j.dld.2004.12.008.

Diels AMJ, Michiels CW. High-pressure homogenization as a non-thermal technique for the inactivation of microorganisms. Critical Rev Microbiol. 2016;32:201–216. doi:https://doi.org/10.1080/10408410601023516.

Dietert RR, Dietert JM. The microbiome and sustainable healthcare. Healthcare. 2015;3:100–129. doi:https://doi.org/10.3390/healthcare3010100.

Donaldson GP, Lee SM, Mazmanian SK. Gut biogeography of the bacterial microbiota. Nat Rev Microbiol. 2016;14(1):20–32. doi:https://doi.org/10.1038/nrmicro3552.

Dore J, Simren M, Buttle L, Guarner F. Hot topics in gut microbiota. United Europ Gastroenterol J. 2013;1(5):311–318. doi:https://doi.org/10.1177/2050640613502477.

Doronin AF, Shenderov BA. Functional nutrition. Moscow: «Grant»; 2002 (in Russian).

Dorrestein PC, Mazmanian SK, Knight R. Finding the missing links among metabolites, microbes, and the host. Immunity. 2014;40:824–832. doi:https://doi.org/10.1016/j.immuni.2014.05.015.

Dzutsev A, Badger JH, Perez-Chanone E, Roy S, Salcedo R et al. Microbes and Cancer. Annu Rev Immunol. 2017;35:199–228. doi:https://doi.org/10.1146/annurev-immunol-051116-052133.

Efaq AN, Rahman NNNA, Nagao H, Al-Gheethi AA, Shahadat M, Kadir MOA. Supercritical carbon dioxide as non-thermal alternative technology for safe handling of clinical wastes. Environmental Processes. 2015;2:797–822.

El Aidy S, Stilling RM, Dinan TG, Cryan JT. Microbiome to Brain: Unravelling the multidirectional axes of communication. Adv Exp Med Biol. 2016;874:301–336.

Emara MH, Mohamed SY, Abdel-Aziz HR. Lactobacillus reuteri in management of Helicobacter pylori infection in dyspeptic patients: a double-blind placebo-controlled randomized clinical trial. Therap Adv Gastroenterol. 2014;7(1):4–13. doi:https://doi.org/10.1177/1756283X13503514.

Engevik MA, Versalovic J. Biochemical feature of beneficial microbes: foundation for therapeutic microbiology. Microbial Spectr. 2017;5(5). doi:https://doi.org/10.1128/microbiolspec.BAD-0012-2016.

Erofeev NP, Radchenko VG, Seliverstov VP. Clinical physiology of the colon. Mechanisms of action of short-chain fatty acids in normal and pathological conditions. St. Petersburg: Forte Print; 2012 (in Russian).

Faith JJ, Ahern PP, Ridaura VK, Cheng J, Gordon JI. Identifying gut microbe-host phenotype relationships using combinatorial communities in gnotobiotic mice. Science Transl Med. 2014;6(220):220ra11. doi:https://doi.org/10.1126/scitranslmed.3008051

Faith JJ, Guruge JL, Charbonneau M, Subramanian S, Seedorf H et al. The long-term stability of the human gut microbiota. Science. 2013;341(6141):1237439. doi:https://doi.org/10.1126/science.1237439.

Falony G, Joossens M, Vieira-Silva S, Wang J, Darzi Y et al. Population-level analysis of gut microbiome variation. Science. 2016;352(6285):560–564. doi:https://doi.org/10.1126/science.aad3503.

Feil R, Fraga MF. Epigenetics and the environment: emerging patterns and implications. Nat Rev Genet. 2012;13:97–109. doi:https://doi.org/10.1038/nrg3142.

Forbes JD, Van Domselaar G, Bernstein CN. The Gut Microbiota in Immune-Mediated Inflammatory Diseases. Front Microbiol. 2016;7:1081. doi:https://doi.org/10.3389/fmicb.2016.01081.

Foster JA, Lyte M, Meyer E, Cryan JF. Gut Microbiota and Brain Function: An Evolving Field in Neuroscience. Int J Neuropsychopharmacol. 2016;19(5):pyv114. doi:https://doi.org/10.1093/ijnp/pyv114.

Francavilla R, Lionetti E, Castellaneta S, Magistà A, Maurogiovanni G et al. Inhibition of Helicobacter pylori infection in humans by Lactobacillus reuteri ATCC 55730 and effect on eradication therapy: a pilot study. Helicobacter. 2008;13:127–134. doi:https://doi.org/10.1111/j.1523-5378.2008.00593.x.

Frank DN, Zhu W, Sartor RB, Li E. Investigating the biological and clinical significance of human dysbioses. Trends Microbiol. 2011;19(9):427–434. doi:https://doi.org/10.1016/j.tim.2011.06.005.

Franzosa EA, Huang K, Meadow JF, Gevers D, Lemon KP et al. Identifying personal microbiomes using metagenomic codes. Peoc Natl Acad Sci. 2015;112:2930–8. doi:https://doi.org/10.1073/pnas.1423854112.

Freitas M, Tavan E, Cayuela C et al. Host-pathogens cross-talk. Indigenous bacteria and probiotics also play the game. Biol Cell. 2003;95(8):503–506.

Frequently Asked Questions About Medical Foods; Second Edition. Guidance for industry. U.S. Dept. Health and Human Services. Food and Drug Administration. Center for Food Safety and Applied Nutrition. 2016. https://www.fda.gov/media/97726/download. Accessed 28 June 2019.

Freter R, Nader de Macias ME. Factors affecting the colonization of the gut by lactobacilli and other bacteria. In: Fuller R et al., editors. Probiotics. Prospects of use in opportunistic infections. Germany: Old Herborn University Seminar; 1995.

Galland L. The gut microbiome and the brain. J Med Food. 2014;17(12):1261–1272.

Ganesh BP, Versalovic J. Luminal conversion and immunoregulation by probiotics. Front Pharmacol. 2015;6:269. doi:https://doi.org/10.3389/fphar.2015.00269.

Gao D, Gao Z, Zhu G. Antioxidant effects of Lactobacillus plantarum via activation of transcription factor Nrf2. Food Funct. 2013;4:982–989.

García-Cañaveras JC, Donato MT, Castell JV, Lahoz A. Targeted profiling of circulating and hepatic bile acids in human, mouse, and rat using a UPLC-MRM-MS-validated method. J Lipid Res. 2012;53:2231–2241. doi:https://doi.org/10.1194/jlr.D028803.

Ghisian Y, Nana G, Ripoll C, Cabin-Flaman A, Gibouin D, Delaune A et al. Division-Based, Growth Rate Diversity in bacteria. Front Microbiol. 2018;9;849. doi:https://doi.org/10.3389/fmicb.2018.00849.

Gibson GR, Roberfroid MB. Dietary modulation of the human colonic microbiota: introducing the concept of prebiotics. J Nutr. 1995;125:1401–1412. doi:https://doi.org/10.1093/jn/125.6.1401.

Gilbert JA, Quinn RA, Debelius J, Morton J, Garg N et al. Microbiome-wide association studies link dynamic microbial consortia to disease. Nature. 2016;535:94–103. doi:https://doi.org/10.1038/nature18850.

Glushanova NA, Shenderov BA. The relationship between probiotic and indigenous lactobacilli in co-culturing in vitro. Zh Mikrobiol (Moscow). 2005;2:75–79 (in Russian).

Golovenko OV, Khalif AO, Golovenko AO. The role of butyric acid in the treatment of organic and functional diseases of the colon. Clinical prospects of gastroenterology, hepatology. 2011;3:20–29 (in Russian).

Green JM, Barratt MJ, Kinch M, Gordon JI. Food and microbiota in the FDA regulatory framework. Science. 2017;357(6346):39–40. doi:https://doi.org/10.1126/science.aan0836.

Gu Q, Li P. Biosynthesis of vitamins by probiotic bacteria. 2016. doi:https://doi.org/10.5772/63117.

Guilherme L, Kallil J, Cunningham M. Molecular mimicry in the autoimmune pathogenesis of rheumatic heart disease. Autoimmunity. 2006;39:31–39. doi:https://doi.org/10.1080/08916930500484674.

Hall A, Versalovic J. Microbial metabolism in the mammalian gut: molecular mechanisms and clinical implications. J Pediatr Gastroenterol Nutr. 2018;66 Suppl 3:72–79. doi:https://doi.org/10.1097/MPG.0000000000001857.

Hanahan D, Weinberg RA. Hallmarks of cancer: the next generation. Cell. 2011;144(5):646–674. doi:https://doi.org/10.1016/j.cell.2011.02.013.

Helminen A, Nummi K, Salaspuro M, Vakevainen S. High concentration of carcinogenic acetaldehyde in fermented dairy products with and without probiotics. In: Proceeding of the International Scientific Conference on Probiotics and Prebiotics. Kosice, Slovakia. 14–16 June 2011:89–90.

Holmes E, Li JV, Athanasiou T, Ashrafian H, Nicholson JK. Understanding the role of gut microbiome-host metabolic signal disruption in health and disease. Trends Microbiol. 2011;19:349–359. doi:https://doi.org/10.1016/j.tim.2011.05.006.

Holz C, Busjahn A, Mehling H, Arya S, Boettner M, Habibi H, Lang C. Significant Reduction in Helicobacter pylori Load in Humans with Non-viable Lactobacillus reuteri. Probiotics Antimicrob Proteins. 2015;7(2):91–100. doi:https://doi.org/10.1007/s12602-014-9181-3.

Hood L. Tackling the microbiome. Science. 2012;336(6086):1209. doi:https://doi.org/10.1126/science.1225475.

Hunt RH, Xiao SD, Megraud F, Leon-Barua R, Bazzoli F et al. Helicobacter pylori in developing countries. World Gastroenterology Organisation Global Guideline. J Gastrointestin Liver Dis. 2011;20(3):299–304.

Huyckel MM, Abrams V, Moore DR. Enterococcus faecalis produces extracellular superoxide and hydrogen peroxide that damage colonic epithelial cell DNA. Carcinogenesis. 2003;23:529–536. doi:https://doi.org/10.1093/carcin/23.3.529.

Ilinskaya ON, Ulyanova VV, Yarullina DR, Gataullin IG. Secretome of intestinal Bacilli; a nature guard against pathologies. Front Microbiol. 2017;8:1666. doi:https://doi.org/10.3389/fmicb.2017.01666.

Indriyani A, Juffrie M, Satyati A. Effect of live verses heat killed probiotics on acute diarrhea in young children. Paediatrica Indonesiana. 2012;52:249–254.

Ishikawa K, Handa N, Sears L, Raleigh EA, Kobayashi I. Cleavage of a model DNA replication fork by a methyl-specific endonuclease. Nucl Acids Res. 2011;39:5489–5498. doi:https://doi.org/10.1093/nar/gkr153.

Jimenez-Chillaron JC, Diaz R, Martinez D, Pentinat T, Ramon-Krauel M, Ribo S, Plosch T. The role of nutrition on epigenetic modifications and their implications on health. Biochimie. 2012;94:2242–2263. doi:https://doi.org/10.1016/j.biochi.2012.06.012.

Jones RM, Neish AS. Redox signaling mediated by the gut microbiota. Free Radical Biology and Medicine. 2017;105:41–47. doi:https://doi.org/10.1016/j.freeradbiomed.2016.10.495.

Jurk D, Wilson C, Passos JF, Oakley F, Correia-Melo C et al. Chronic inflammation induces telomere dysfunction and accelerates ageing in mice. Nat Commun. 2014;2:4172. doi:https://doi.org/10.1038/ncomms5172.

Kamada N, Nunez G. Regulation of the immune system by the resident intestinal bacteria. Gastroenterology. 2014;146(6):1477–1488. doi:https://doi.org/10.1053/j.gastro.2014.01.060.

Kanherkar RR, Bhatia-Dey N, Makarev E, Csoka AB. Cellular reprogramming for understanding and treating human disease. Frontiers in Cell and Development biology. 2014a;2. doi:https://doi.org/10.3389/fcell.2014.00067.

Kanherkar RR, Bhatia-Dey N, Csoka AB. Epigenetics across the human lifespan. Front Cell Dev Biol. 2014b;2:49. doi: https://doi.org/10.3389/fcell.2014.00049.

Karczewski J, Poniedzialek B, Adamski Z, Rzymski P. The effects of the microbiota on the host immune system. Autoimmunity. 2014;47(8):494–504. doi:https://doi.org/10.3109/08916934.2014.938322.

Karczewski KJ, Snyder MP. Integrative omics for health and disease. Nature Reviews Genetics. 2018;19(5):299–309. doi:https://doi.org/10.1038/nrg.2018.4.

Katan MB. Why the European Food Safety Authority was right to reject health claims or probiotics. Benef Microb. 2012;3(2):85–89. doi:https://doi.org/10.3920/BM2012.0008.

Kazerouni A, Burgess J, Burns LJ, Wein LM. Optimal screening and donor management in a public stool bank. Microbiome. 2015;3:75. doi:https://doi.org/10.1186/s40168-015-0140-3.

Khoder G, Al-Menhali AA, Al-Yassir F, Karam SM. Potential role of probiotics in the management of gastric ulcer. Exp Ther Med. 2016;12(1):3–17. doi:https://doi.org/10.3892/etm.2016.3293.

Kim Y, Edwards N, Fenselau C. Extracellular vesicle proteomes reflect developmental phases of Bacillus subtilis. Clin Proteomics. 2016;13:6. doi:https://doi.org/10.1186/s12014-016-9107-z.

Kinnebrew MA, Buffi CG, Diehl GE, Zenewicz LA, Leiner I et al. Interleukin 23 production by intestinal CD103(+)CD11b(+) dendritic cells in response to bacterial flagellin enhances mucosal innate immune defense. Immunity. 2012;36(2):276–287. doi:https://doi.org/10.1016/j.immuni.2011.12.011.

Kiseleva EP, Mikhailopulo KI, Svitidov OV, Novik GI, Knirel YA, Dey ES. The role of components of bifidobacterium and Lactobacillus in pathogenesis and serologic diagnosis of autoimmune thyroid diseases. Benef Microb. 2011;2(2):139–154. doi:https://doi.org/10.3920/BM2010.0011.

Klein G, Schanstra JP, Hoffmann J, Mischak H, Siwy J, Zimmerman K. Proteomics as a quality control tool of pharmaceutical probiotic bacterial lysate products. PLoS One. 2013;8(6):e66682. doi:https://doi.org/10.1371/journal.pone.0066682.

Kocharovec VI. Hilak Forte: modern model metabiotics with biotherapeutic activity. Moscow: Prime-Media; 2018 (in Russian).

Lachnit T, Bosch TCG, Deines P. Exposure of the Host-Associated Microbiome to Nutrient-Rich Conditions May Lead to Dysbiosis and Disease Development — an Evolutionary Perspective. mBio. 2019;10(3):e00355–19. doi:https://doi.org/10.1128/mBio.00355-19.

Lagier JC, Million M, Hugon P, Armougom F, Raoult D. Human gut microbiota: repertoire and variations. Front Cell Infect Microbiol. 2012;2:136. doi:https://doi.org/10.3389/fcimb.2012.00136.

Laino J, Villena J, Kanmani P, Kitazawa H. Immunoregulatory effects triggered by lactic acid bacteria exopolysaccharides: new insights into molecular interaction with host cells. Microorganisms. 2016;4:27. doi:https://doi.org/10.3390/microorganisms4030027.

Lakhtin V, Lakhtin MV, Pospelova VV, Shenderov BA. Lectins of lactobacilli and bifidobacteria. II Probiotic lectins of lactobacilli and bifidobacteria as possible signal molecules regulating inter- and intrapopulation relationships between bacteria and between bacteria and the host. Microb Ecol Health Dis. 2007;19(3):153–157. doi:https://doi.org/10.1080/08910600701538257.

Lakhtin VM, Lakhtin MV, Cherepanova YuV, Pospelova VV, Afanasyev SS, Aleshkin VA. High molecular weight biosurfactants of human gram-positive bacteria. Clin lab diagn. 2010;9:37–38 (in Russian).

Lebeer S, Bron PA, Marco ML, Van Pijkeren JP, O'Conell Motherway M et al. Identification of probiotic effect or molecules: present state and future perspectives. Curr Opin Biotechnol. 2018;49:217–223. doi:https://doi.org/10.1016/j.copbio.2017.10.007.

Lebeer S, Vanderleyden J, De Keersmaeker SCJ. Genes and molecules of lactobacilli supporting probiotic action. Microbiol Mol Biol Rev. 2008;72(4):728–764. doi:https://doi.org/10.1128/MMBR.00017-08.

Lebeer S, Vanderleyden J, De Keersmaeker SCJ. Host interactions of probiotic bacterial surface molecules: comparison with commensals and pathogens. Natl Rev Microbiol. 2010; 8(3):171–184. doi:https://doi.org/10.1038/nrmicro2297.

Lederberg J, McCray A.T. 'Ome Sweet' Omics — a genealogical treasury of words. Scientist. 2001;15(7): 8.

Lederberg J. Infectious history. Science. 2000;288(5464):287–293. doi:https://doi.org/10.1126/science.288.5464.287.

Lee MJ, Zang Z, Choi EY, Shin HK, Ji GE. Cytoskeleton reorganization and cytokine production of macrophages by Bifidobacterial cells and cell-free extracts. Journal of Microbiology and Biotechnology. 2002;12(3):398–405.

Lefevre M, Racedo SM, Ripert G, Housez B, Cazaubiel M et al. Probiotic strain Bacillus subtilis CU1 stimulates immune system of elderly during common infectious disease period: a randomized, double-blind placebo-controlled study. Immun Ageing. 2015;12:24. doi:https://doi.org/10.1186/s12979-015-0051-y.

Legent G, Norris V. Hybolites: novel therapeutic toolsfor targeting hyperstructure in bacteria. Recent Pat on Antiinfect Drug Discov. 2009;4(2):90–95.

Lenoir-Wijnkoop I, Sanders ME, Cabana MD, Van Loo J, Sherman PM et al. Probiotic and probiotic influence beyond the intestinal tract. Nutr Rev. 2007;65(11):469–489. doi:https://doi.org/10.1111/j.1753-4887.2007.tb00272.x.

Lesbros-Pantoflickova D, Corthesy-Theulaz I, Blum AL. Helicobacter pylori and probiotics. J Nutr. 2007;137(3 Suppl 2):812S–818S. doi:https://doi.org/10.1093/jn/137.3.812S.

Li M, Wang B, Zhang M, Rantalainen M, Wang S et al. Symbiotic gut microbes modulate human metabolic phenotypes. Proc Natl Avad Sci USA. 2008;105(6):2117–2122. doi:https://doi.org/10.1073/pnas.0712038105.

Li Y, Tollefsbol TO. Impact on DNA methylation in cancer prevention and therapy by bioactive compounds. Curr Med Chem. 2010; 17(20):2141–2151.

Linsalata M, Russo F, Berloco P, Caruso ML, Matteo GD, Cifone MG, Simone CD, Ierardi E, Di Leo A. The influence of Lactobacillus brevis on ornithine decarboxylase activity and polyamine profiles in Helicobacter pylori-infected gastric mucosa. Helicobacter 2004;9:165–172. doi:https://doi.org/10.1111/j.1083-4389.2004.00214.x.

Lionetti E, Miniello V, Castellaneta S, Magistá A, de Canio A, Maurogiovanni G. et al. Lactobacillus reuteri therapy to reduce side-effects during anti-Helicobacter pylori treatment in children: a randomized placebo controlled trial, Aliment Pharmacol Ther 2006;24:1461–1468. doi:https://doi.org/10.1111/j.1365-2036.2006.03145.x.

López-Otín C, Blasco MA, Partridge L, Serrano M, Kroemer G. The hallmarks of aging. Cell. 2013;153(6):1194–1217. doi:https://doi.org/10.1016/j.cell.2013.05.039.

Lyte M. Microbial endocrinology and nutrition: A perspective on new mechanisms by which diet can influence gut-to brain-communication. Pharma Nutrition. 2013;1(1):35–39.doi:https://doi.org/10.1016/j.phanu.2012.11.002.

Lyu Q, Hsu C-C. Can diet influence our health by altering intestinal microbiota-derived fecal metabolites? mSystems. 2018;3(2):e00187–17. doi:https://doi.org/10.1128/mSystems.00187-17.

Maguire M, Maguire G. Gut dysbiosis, leaky gut, and intestinal epithelial proliferation in neurological disorders: towards the development of a new therapeutic using amino acids, prebiotics, probiotics, and postbiotics. Rev Neurosci. 2019;30(2):179–201. doi:https://doi.org/10.1515/revneuro-2018-0024.

Malfertheiner P, Megraud F, O'Morain CA, Gisbert JP, Kuipers EJ et al. Management of Helicobacter pylori infection-the Maastricht V / Florence Consensus Report. Gut. 2017;66(1):6–30. doi:https://doi.org/10.1136/gutjnl-2016-312288.

Marco ML, Pavan S, Kleerebezem M. Towards understanding molecular modes of probiotic action. Curr Opin Biotechnol. 2006;17(2):204–210. doi:https://doi.org/10.1016/j.copbio.2006.02.005.

Marcobal A, Kashyap P.C, Nelson T.A, Aronov P.A, Donia M.S, Spormann A, Fischbach M.A, Sonnenburg J.L. A metabolomic view of how the human gut microbiota impacts the host metabolome using humanized and gnotobiotic mice. ISME J. 2013;7(10):1933–1943. doi:https://doi.org/10.1038/ismej.2013.89.

Martin FPJ, Wang Y, Sprenger N, Yap IK, Lundstedt T et al. Probiotic modulation of symbiotic gut microbial-host metabolic interactions ib a humanized microbiome mouse model. Molecular Systems Biology. 2008;4:157. doi:https://doi.org/10.1038/msb4100190.

Matsumoto K, Hara H, Fishov I, Mileykovskaya E, Norris V. The membrane: transertion as an organizing principle in membrane heterogeneity. Front Microbiology. 2015;6:572. doi:https://doi.org/10.3389/fmicb.2015.00572.

Mehling H, Busjahn A. Non-viable Lactobacillus reuteri DSMZ17648 (Pylopass™) as a new approach to Helicobacter pylori control in humans. Nutrients. 2013;5(8):3062–3073. doi:https://doi.org/10.3390/nu5083062.

Meijer K, de Vos P, Priebe MG. Butyrate and other short-chain fatty acids as modulators of immunity: what relevance foe health. Curr Opin Clin Nutr Metab Care. 2010;13(6):715–21. doi:https://doi.org/10.1097/MCO.0b013e32833eebe5.

Meisel JC, Grice EA. The human microbiome. Chapter 4. In: Ginsburg G, Willard H, editors. Genomic and Precision Medicine (Third Edition). Elsevier Inc; 2017. p. 63–77. doi:https://doi.org/10.1016/B978-0-12-800681-8.00004-9.

Michetti P, Dorta G, Wiesel PH, Brassart D, Verdu E, Herranz M, Felley C, Porta N, Rouvet M, Blum AL, Corthésy-Theulaz I. Effect of whey-based culture supernatant of Lactobacillus acidophilus (johnsonii) La1 on Helicobacter pylori infection in humans. Digestion 1999;60:203–209. doi:https://doi.org/10.1159/000007660.

Midtvedt T. Microflora-associated characteristics (MACs) and germfree animal characteristics (GACs) in man and animal. Microecology and Therapy. 1985;15:295–302.

Mimee M, Citorik RJ, Lu TK. Microbiome therapeutics-Advances and challenges. Adv Drug Deliv Rev. 2016;105(Pt A):44–54. doi:https://doi.org/10.1016/j.addr.2016.04.032.

Minushkin ON, Ardatskaya MD, Sergeev AV, Volkov MYu, Sinitsa AV. Experience in the use of probiotic «Baktistatin» in the treatment of chronic pancreatitis. Pharmacy. 2006;3:39–43 (in Russian).

Mischke M, Plosch T. The gut microbiota and their metabolites: Potential implications for the host epigenome. Adv Exp Med Biol. 2016;902:33–44. doi:https://doi.org/10.1007/978-3-319-31248-4_3.

Mishra V, Shah C, Mokashe N, Chavan R, Yadav H, Prajapati J. Probiotics as potential antioxidants: A systematic review. J. Agric Food Chem. 2015;63(14):3615–3626. doi:https://doi.org/10.1021/jf506326t.

Moreira MM, Morais S, Delerue-Matos C. Environment-Friendly Techniques for Extraction of Bioactive Compounds From Fruit. Chapter 2. In: Handbook of Food Bioengineering. Grumezescu AM, Holban AM, editors. Vol. 3. Soft Chemistry and Food Fermentation. London: Elsevier Inc.; 2017. p. 21–77. doi:https://doi.org/10.1016/B978-0-12-811412-4.00002-3.

Mukai T, Asasaka T, Sato E, Mori K, Matsumoto M, Ohori H. Inhibition of binding of Helicobacter pylori to the glycolipid receptors by probiotic Lactobacillus reuteri. FEMS Immunol Med Microbiol. 2002;32(2):105–10. doi:https://doi.org/10.1111/j.1574-695X.2002.tb00541.x.

Nakagawa H, Miyazaki T. Anti-aging effects of Lactobacilli. Integr Mol Med. 2016;3(3):680–681. doi:https://doi.org/10.15761/IMM.1000223.

Nakamura Y, Bando I, Mennear JY, Bernard BK. Studies of toxicological potential of tripeptides. IV. Assessment of the repeated-dose toxicological potential of synthesized L-valyl-L-prolyl-L-proline in male and female rats and dogs. Int J Toxicol. 2005;24:25–39.

Nakamura Y. Studies on anti-hypertensive peptides in milk fermented with *Lactobacillus helveticus*. Biosci Microflora. 2004;23:131–138. doi:https://doi.org/10.12938/bifidus.23.131.

Neis EP, Dejong CH, Rensen SS. The role of microbial amino acids metabolism. Nutrients. 2015;7(4):2930–2946. doi:https://doi.org/10.3390/nu7042930.

Neish AS. Microbes in gastrointestinal health and disease. Gastroenterology. 2009;136(1):65–80. doi:https://doi.org/10.1053/j.gastro.2008.10.080.

Ng SC, Harf AL, Kamm MA, Stagg AJ, Knight SC. Mechanism of action on probiotics: recent advances. Inflamm Bowel Dis. 2009;15(2):300–310. doi:https://doi.org/10.1002/ibd.20602.

Nicholson JK, Holmes E, Kinross J, Gibson G, Jia W, Pettersson S. Host-Gut microbiota metabolic interactions. Science. 2012;336(6086):1262–1267. doi:https://doi.org/10.1126/science.1223813.

Nicholson JK, Holmes E, Wilson ID. Gut microorganisms mammalian metabolism and personalized health care. Nat Rev Microbiol. 2005;3:431–438. doi:https://doi.org/10.1038/nrmicro1152.

Nicolson GL, de Mattos GF, Settineri R, Costa C, Ellithorpe R et al. Clinical effects of hydrogen administration: from animal and human diseases to exercise medicine. Intern J Clin Medicine. 2016;7(1):32–76. doi:https://doi.org/10.4236/ijcm.2016.71005.

Norris V, Verrier C, Feuilloley M. Hybolites Revisited. Recent Pat on Anti-Infect Drug Discov. 2016;11(1):1–13.

Nougayrede JP, Homburg S, Taieb F, Boury M, Brzuszkiewicz E, Gottschalk G et al. Escherichia coli induces DNA double-strand breaks in eukaryotic cells. Science 2006;313(5788):848–851. doi:https://doi.org/10.1126/science.1127059.

Nummi KP, Helminen PA, Salaspuro M, Vakevainen S. Ethanol concentration of probiotic and basic dairy products — a potential risk for upper digestive tract cancer. In: Proceeding of the International Scientific Conference on Probiotics and Prebiotics, Kosice, Slovakia. 14–16 June 2011. P.107–8.

O'Flaherty S, Kleenhammer TR. The impact of omic technologies on the study of food microbes. Annu Rev Food Sci Technol. 2011;2:353–371. doi:https://doi.org/10.1146/annurev-food-030810-110338.

O'Callaghan TF, Ross RP, Stanton C, Clarke G. The gut microbiome as a virtual endocrine organ with implications for farm and domestic animal endocrinology. Domest Anim Endocrinol. 2016;56 Suppl:S44–55. doi:https://doi.org/10.1016/j.domaniend.2016.05.003.

Ohishi A, Takahashi S, Ito Y, Ohishi Y, Tsukamoto K et al. Bifidobacterium septicemia associated with postoperative probiotic therapy in a neonate with omphalocele. J Pediatrics. 2010;156(4):679–681. doi:https://doi.org/10.1016/j.jpeds.2009.11.041.

Oleskin AV, El'-Registan GI, Shenderov BA. Role of neuromediators in the functioning of the human microbiota: «business talks» among microorganisms and the microbiota-host dialogue. Microbiology. 2016;85(1):1–22.

Oleskin AV, Shenderov BA, Rogovsky VS. Role of Neurochemicals in the Interaction between the Microbiota and the Immune and the Nervous System of the Host Organism. Probiotics and Antimicrobial Proteins. 2017;9(3):215–234. doi:https://doi.org/10.1007/s12602-017-9262-1.

Oleskin AV, Shenderov BA. Neuromodulatory effects and targets of the SCFAs and gasotransmitters produced by the human symbiotic microbiota. Microb Ecol Health Dis. 2016;27:30971. doi:https://doi.org/10.3402/mehd.v27.30971.

Oleskin AV, Zhilenkova OG, Shenderov BA, Adelaide M, Amerhanova AM, Kudrin VS, Klodt PM. Lactic-Acid Bacteria Supplement Fermented Dairy Products with Human Behavior-Modifying Neuroactive Compounds. Jour Pharm Nutr Sci. 2014;4:199–206. doi:https://doi.org/10.6000/1927-5951.2014.04.03.5.

Olmos J, Paniagua-Michel J. *Bacillus subtilis* A Potential Probiotic Bacterium to Formulate Functional Feeds for Aquaculture. J Microb Biochem Technol. 2014;6:361–365. doi:https://doi.org/10.4172/1948-5948.1000169.

Osipov GA, Verkhovtseva NV. Study of human microecology by mass-spectrometry of microbial markers. Benefic Microbes. 2011;2(1):63–78. doi:https://doi.org/10.3920/BM2010.0017.

Pang IK, Iwasaki A. Control of antiviral immunity by pattern recognition and the microbiome. Immunol Rev. 2012;245(1):209–226. doi:https://doi.org/10.1111/j.1600-065X.2011.01073.x.

Panyushin SK. Participation of symbiotic microflora in providing energy needs of the person. 2012. www. biophys.ru/archive/congress2012/proc-p273.htm. Accessed 28 June 2019 (in Russian).

Paul B, Barnes S, Demark-Wahnefried W, Morrow C, Salvador C et al. Influences of diet and the gut microbiome on epigenetic modulation in cancer and other diseases. Clinical Epigenetics. 2015;7:112. doi:https://doi.org/10.1186/s13148-015-0144-7.

Perez KJ, Viana JD, Lopes FC, Pereira JQ, Dos Santos DM et al. Bacillus spp isolated from puba as a source of biosurfactants and antimicrobial lipopeptides. Front Microbiol. 2017;8:61. doi:https://doi.org/10.3389/fmicb.2017.00061.

Pérez-Jiménez J, Fezeu L, Touvier M et al. Dietary intake of 337 polyphenols in French adults. Am J Clin Nutr. 2011;93(6):1220–1228. doi:https://doi.org/10.3945/ajcn.110.007096.

Pflughoeft KJ, Versalovic J. Human microbiome in health and disease. Ann Rev Pathol Mech Dis. 2012;7:99–122. doi:https://doi.org/10.1146/annurev-pathol-011811-132421.

Phister TG, O'Sullivan DJ, McKay LL. Identification of bacilysin, chlorotetaine, and iturin a produced by Bacillus sp. strain CS93 isolated from Pozol, a Mexican fermented maize dough. Appl Environm Microbiol. 2004;70(1):631–634. doi:https://doi.org/10.1128/AEM.70.1.631-634.2004.

Ploskireva AA. Metabolic therapy of microbiocenosis disorders in different biotopes of the human body. Attending Physician. 2016;7(2):632–642 (in Russian).

Popova TS, Zhuravel SV, Chernenkaya TV, Tropskaya NS. Preclinical and clinical evaluation of the effectiveness of the product for functional nutrition «Extract of lactic acid bacteria fermented in soy milk «Daigo». Moscow: Daigo; 2016 (in Russian).

Prangli AL, Utt M, Talja I, Sepp E, Mikelsaar M, Rajasalu T et al. Antigenic proteins of lactobacillus acidophilus that are recognized by serum IgG antibodies in children with type I diabetes and celiac diseases. Pediatr Allergy Immunol. 2010;21(4 Pt 2):e772–779. doi:https://doi.org/10.1111/j.1399-3038.2009.00904.x

Proal AD, Lindseth IA, Marshall TG. Microbe-Microbe and Host-Microbe Interactions Drive Microbiome Dysbiosis and Inflammatory Processes. Discovery Medicine. 2017;23(124):51–60.

Prosekov A, Dyshlyuk L, Milentyeva I, Sukhish S, Babich O et al. Antioxidant, antimicrobial and antitumor activity of bacteria of the genus Bifidobacterium, selected from the gastrointestinal tract of human. Integr Mol Med. 2015;2(5):295–303.

Quartieri A, García-Villalba R, Amaretti A et al. Detection of novel metabolites of flaxseed lignans in vitro and in vivo. Mol Nutr Food Res. 2016;60(7):1590–1601. doi:https://doi.org/10.1002/mnfr.201500773.

Quintana VM, Torres NI, Wachsman MB, Sinko PJ, Castilla V, Chikindas M. Antiherpes simplex virus type 2 activity of the antimicrobial peptide subtilosin. J Appl Microbiol. 2014;117(5):1253–1259. doi:https://doi.org/10.1111/jam.12618.

Rajilic-Stojanovic M, de Vos WM. The first 1000 cultured species of the human gastrointestinal microbiota. FEMS Microbiol Rev. 2014;38:996–1047. doi:https://doi.org/10.1111/1574-6976.12075.

Reading NC, Sperandio V. Quorum sensing: the many languages of bacteria. FEMS Microbiol Let. 2006;254(1):1–11. doi:https://doi.org/10.1111/j.1574-6968.2005.00001.x.

Reid G, Younes JA, Mei Van der HC, Gloor GB et al. Microbiota restoration: natural and supplemented recovery of human microbial communities. Nat Rev Microbiol. 2011;9(1):27–38. doi:https://doi.org/10.1038/nrmicro2473.

Remely M, Lovrecic L, de la Garza AL, Migliore L, Peterlin B et al. Therapeutic perspectives of epigenetically active nutrients. British J Pharmacol. 2015;172(11):2756–2768. doi:https://doi.org/10.1111/bph.12854.

Roberfroid M, Gibson GR, Hoyles L, McCartney AL et al. Prebiotic effects: metabolic and health benefits. Br J Nutr. 2010;104 Suppl 2:1–63. doi:https://doi.org/10.1017/S0007114510003363.

Roda A, Simoni P, Magliulo M, Nanni P, Baraidini M et al. A new oral formulation for the release of sodium butyrate in the ileo-cecal region and colon. World J Gastroenterol. 2007;13(7):1079–1084. doi:https://doi.org/10.3748/wjg.v13.i7.1079.

Rudkowski Z, Bromirska J. Shortening of the period of fecal excretion of salmonella in infants under treatment with Hylak forte. Paediatrie und Paedologie. 1991; 26(2):111–114. https://eurekamag.com/research/007/790/007790996.php. .

Russell WR, Duncan SH, Scobbie L, Duncan G, Cantlay L et al. Major phenylpropanoid-derived metabolites in the human gut can arise from microbial fermentation of protein. Mol Nutr Food Res. 2013;57(3):523–535. doi:https://doi.org/10.1002/mnfr.201200594.

Said HM. Recent advances in transport of water-soluble vitamins in organs of the digestive system: a focus on the colon and the pancreas. Am J Physiol Gastrointest Liver Physiol. 2013;305(9):G601–610. doi:https://doi.org/10.1152/ajpgi.00231.2013.

Sampath H, Ntambi JM. Polyunsaturated fatty acid regulation of genes of lipid metabolism. Annu Rev Nutr. 2005;25:317–340. doi:https://doi.org/10.1146/annurev.nutr.25.051804.101917.

Sampson TR, Mazmanitan SK. Control of brain development, function, and behavior by the microbiome. Cell Host Microbe. 2015;17(5):565–576. doi:https://doi.org/10.1016/j.chom.2015.04.011.

Sanders ME, Shaneb AL, Merensteinc DJ. Advancing probiotic research in humans in the United States: Challenges and strategies. Gut Microbes. 2016;7(2):97–100. doi:https://doi.org/10.1080/19490976.2016.1138198.

Sanshez S, Demain AL. Antibiotics: Current Innovations and Future Trends. Caister Academic Press; 2015.

Savidge TC. Epigenetic Regulation of Enteric Neurotransmission by Gut Bacteria. Front Cell Neurosci. 2015;9:503. doi:https://doi.org/10.3389/fncel.2015.00503.

Sawada H, Furushiro M, Hirai K, Motoike M, Watanabe T, Yokokura T. Purification and characterization of an antihypertensive compound from Lactobacillus casei. Agric Biol Chem. 1990;54:3211–3219.

Scaccianoce G, Zullo A, Hassan C, Gentili F, Cristofari F et al. Triple therapies plus different probiotics for Helicobacter pylori eradication. Eur Rev Med Pharmacol Sci. 2008;12(4):251–256.

Schauder S, Bassler BL. The languages of bacteria. Genes Dev. 2001;15(12):1468–1480.

Sebastian AP, Keerthi TR. Immunomodulatory effect of probiotic strain Bacillus subtilis MBTU PBBM1 spores in Balb/C mice. Int J Eng Tech Res. 2014;2(11):258–260.

Sender R, Fuchs S, Milo R. Are We Really Vastly Outnumbered? Revisiting the Ratio of Bacterial to Host Cells in Humans. Cell. 2016;164(3):337–340. doi:https://doi.org/10.1016/j.cell.2016.01.013.

Shaikh AM, Sreeja V. Metabiotics and their Health Benefits. Intl. J. Food. Ferment. 2017;6(1):11–23. doi:https://doi.org/10.5958/2321-712X.2017.00002.3.

Sharma M, Shukla G. Metabiotics: One Step ahead of Probiotics; an Insight into Mechanisms Involved in Anticancerous Effect in Colorectal Cancer. Front. Microbiol. 2016;7:1940. doi:https://doi.org/10.3389/fmicb.2016.01940.

Shen ZH, Zhu CX, Quan YS, Yang ZY, Wu S et al. Relationship between intestinal microbiota and ulcerative colitis: Mechanisms and clinical application of probiotics and fecal microbiota transplantation. World J Gastroenterol. 2018;24(1):5–14. doi:https://doi.org/10.3748/wjg.v24.i1.5

Shenderov B, Shevyreva M, Makarov V, Zagainova A, Yudin S. Federal cryobank of natural symbiotic microbiocenoses of human, animals, plants, soils and open reservoirs of the Russian Federation. In: Conference proceedings XXXVII Annual Meeting of the European Culture Collections' Organization. 13–15 September 2018. Moscow, Russia; 2018c. p. 101–102.

Shenderov BA, Aleshkin VA. Functional Dairy Foods in Human Epigenetic Regulation. In: Abstract Book IDF WORLD DAIRY SUMMIT. 4–8 November 2012. Cape Town. Poster 0121.

Shenderov BA, Gakhova EN, Kaurova SA, Uteshev VK, Shishova NV. Cryobanks of natural symbiotic microbiocenosis and their importance in medicine and biotechnology. Biophysics of live cell. 2014;10:221–223 (in Russian).

Shenderov BA, Glushanova NA. Experimental substantiation of new approaches to intestinal microbial ecology correction by means of probiotics. Microb Ecol Health Dis. 2006;18:93–94. doi:https://doi.org/10.1080/08910600601056699.

Shenderov BA, Ivanova YaV, Sorokina IM. Complex food product enrichment. PATENT RF 2397246. 2010 (in Russian).

Shenderov BA, Lakhtin MV. Lectins as a new potential category of physiologically active functional food ingredients. Journal of Restorative Medicine & Rehabilitation. 2004;1:33–38 (in Russian).

Shenderov BA, Manvelova MA, Gahova EN, Piorunski DA. Functional food and cryobanks of microbiocenoses-practical applications of gnotobiology in the 21st century. In: Hashimoto K et al, editors. Germfree life and its ramifications. Japan: Shiozawa; 1996. p. 533–536.

Shenderov BA, Manvelova MA, Stepanchuk JuB, Skiba NE. Probiotic and Functional Nutrition. Antibiotics and Chemotherapy. 1997;42(7):30–34 (in Russian).

Shenderov BA, Midtvedt T. Epigenomic programing: a future way to health? Microb Ecol Health Dis. 2014;25:24145. doi:https://doi.org/10.3402/mehd.v25.24145.

Shenderov BA, Mitrokhin DD, Zaslavskaya PL. The effect of antibiotics on excretion of different metabolites in the faeces of rats. Microecology and Therapy. 1990;20:53–61.

Shenderov BA, Mitrokhin SD, Glukhova EV, Vagina IM, Ivanova LV et al. The influence of antibiotics on excretion with faeces of some microbial metabolites. Antibiotics. 1989;34:665–669.

Shenderov BA, Tkachenko EI, Lazebnik LB, Ardatskaya MD, Sinitsa AV, Zakharchenko MM. Metabiotics-novel technology of protective and treatment of diseases associated with microecological imbalance in human beings. Experimental & Clinical Gastroenterology. 2018a;151:83–92 (in Russian).

Shenderov BA, Yudin SM, Shevyreva MP, Boyko EA. The scientific backgrounds for the creation of a microecological cryopreservation of human resources. Hygiene and Sanitation. 2018b;97(5):396–398 (in Russian).

Shenderov BA. Directions of gnotobiology development: theoretical and practical aspects. Treatment and prevention. 2014a;1:80–84 (in Russian).

Shenderov BA. Functional food, cryobanks of microbial associations and their role in the preservation and restoration of health. Journal of Restorative Medicine & Rehabilitation. 2003;3:29–31 (in Russian).

Shenderov BA. Functional nutrition and its role in the prevention of metabolic syndrome. Moscow: DeLi print; 2008 (in Russian).

Shenderov BA. Medical microbial ecology and functional nutrition. Part 3. In: Shenderov BA. Probiotics and functional nutrition. Moscow: Grant; 2001 (in Russian).

Shenderov BA. Metabiotics — novel prophylactic technology of diseases associated with microecological imbalance of human being. Journal of Restorative Medicine & Rehabilitation. 2017;4:40–49 (in Russian).

Shenderov BA. Metabiotics: novel idea or natural development of probiotic conception. Microb Ecol Health Dis. 2013a;24:20399. doi:https://doi.org/10.3402/mehd.v24i0.20399.

Shenderov BA. Microbial ecology and its role in promoting health. Metamorphoses. 2014b;5:72–80 (in Russian).

Shenderov BA. Microecological epigenetic of stress, diseases, health and longevity. Journal of Restorative Medicine & Rehabilitation. 2016a;1:21–28 (in Russian).

Shenderov BA. Modern Condition and Prospective Host Microecology Investigations. Microb Ecol Heath Dis. 2007;19(3):145–149. doi:https://doi.org/10.1080/08910600701520933.

Shenderov BA. Molecular language of probiotic microorganisms. Food ingredients: Raw materials & additives. 2009;1:47–48 (in Russian).

Shenderov BA. OMIC-technologies and their importance in current prophylactic and regenerative medicine. Journal of Restorative Medicine & Rehabilitation. 2012;3:70–78 (in Russian).

Shenderov BA. Probiotic (symbiotic) bacterial languages. Anaerobe. 2011a;17(6):490–495. doi:https://doi.org/10.1016/j.anaerobe.2011.05.009.

Shenderov BA. Probiotics and Functional Foods. In: Food Engineering, Encyclopedia of Life Support Systems (EOLSS). Developed under the Auspices of the UNESCO. Oxford, UK: EOLSS Publishers; 2011b.

Shenderov BA. Role of mitochondria in preventive, restorative and sport medicine. Journal of Restorative Medicine & Rehabilitation. 2018;1:21–31 (in Russian).

Shenderov BA. Targets and effects of short chain fatty acids. J Mod Med Sci. 2013b;1–2:21–50 (in Russian).

Shenderov BA. The microbiota as an epigenetic control mechanism. Chapter 11. In: Nibali L, Henderson B, editors. The Human Microbiota and Chronic Disease: Dysbiosis as a Cause of Human Pathology. London: John Willey & Sons, Inc.; 2016b. p. 179–197.

Shenderov BA. The role of endogenous and microbial gas molecules in physiology, pathophysiology of cardiovascular system. Journal of Restorative Medicine & Rehabilitation. 2015a;5:58–65 (in Russian).

Shenderov BA. The role of nutrition and intestinal microflora in programming and realization of epigenome of healthy and sick people. Journal of Restorative Medicine & Rehabilitation. 2013c; January, Special issue:102–107 (in Russian).

Shenderov BA. The role of nutrition and symbiotic microbiota in the epigenetics of chronic somatic diseases. Dietetics issues. 2015b;5(1):22–23 (in Russian).

Singh A, Vishwakarma V, Singhal B. Metabiotics: The Functional Metabolic Signatures of Probiotics: Current State-of-Art and Future Research Priorities — Metabiotics: Probiotics Effector Molecules. Advan Biosci Biotechnol. 2018;9(4):147–189. doi:https://doi.org/10.4236/abb.2018.94012.

Sitkin SI, Tkachenko EI, Vakhitov TYa. Phylo-metabolic nucleus of an intestinal microbiota. Almanac of clinical medicine. 2015;40:12–34 (in Russian).

Sobol CV, Sobol YT. Composition and method to produce and use fermented hydrolyzed products of microbial metabolism. Canada Patent 2484639. 2012.

Sobol CV. A new class of Pharmabiotics with Unique Properties. Chapter 4. In: Grumezescu AM, Holban AM, editors. Handbook of Food Bioengineering. Vol. 3. Soft Chemistry and Food Fermentation. London: Elsevier Inc.; 2017. p. 79–108. doi:https://doi.org/10.1016/B978-0-12-811412-4.00004-7.

Sommer MO, Church GM, Dantas G. The human microbiome harbors a diverse reservoir of antibiotic resistance genes. Virulence. 2010;1(4):299–303. doi:https://doi.org/10.4161/viru.1.4.12010.

Sonnenburg JL, Backhed F. Diet-microbiota interactions as moderators of human metabolism. Nature. 2016;535(7610):56–64. doi:https://doi.org/10.1038/nature18846.

Sonnenburg JL, Fischbach MA. Community health care: therapeutic opportunities in the human microbiome. Sci Transl Med. 2011;3(78):78ps12. doi:https://doi.org/10.1126/scitranslmed.3001626.

Sorokulova I. Modern status and perspectives of Bacillus bacteria as Probiotics. J Prob Health. 2013;1(4):e106. doi:https://doi.org/10.4172/2329-8901.1000e106.

Stilling RM, Dinan TG, Cryan JF. Microbial genes, brain & behaviour – epigenetic regulation of the gut-brain axis. Genes Brain Behav. 2014;13(1):69–86. doi:https://doi.org/10.1111/gbb.12109.

Suerbaum S, Michetti P. *Helicobacter pylori* infection. N Engl J Med. 2002;347(15):1175–1186. doi:https://doi.org/10.1056/NEJMra020542.

Sumi CD, Yang BW, Yeo IC, Hahm YT. Antimicrobial peptides of the genus *Bacillus*. A new era for antibiotics. Can J Microbiol. 2015;61(2):93–103. doi:https://doi.org/10.1139/cjm-2014-0613.

Suva MA, Sureja VP, Kheni DB. Novel insight on probiotic *Bacillus subtilis*: Mechanism of action and clinical applications. J Curr Res Sci Med. 2016;2:65–72. doi:https://doi.org/10.4103/2455-3069.198381.

Suvorov A, Karaseva A, Kotyleva M, Kondratenko Y, Lavrenova H et al. Autoprobiotics as an approach for restoration of personalized microbiota. Front Microbiol. 2018;9:1869. doi:https://doi.org/10.3389/fmicb.2018.01869.

Suvorov A. Gut microbiota, probiotics, and human health. Bioscience of Microbiota, Food and Health. 2013;32(3):81–91. doi:https://doi.org/10.12938/bmfh.32.81.

Taga ME, Bassler BL. Chemical communication among bacteria. Proc Natl Acad Sci USA. 2003;100 Suppl 2:14549–14554. doi:https://doi.org/10.1073/pnas.1934514100.

Tamm AO, Viya MP, Mikelsaar ME, Sijgur UKh. Metabolites of the intestinal flora in the diagnostic of dysbiosis of the intestines. Antibiotics. 1987;32:191–195.

Tan DX, Manchester LC, Esteban-Zubero E, Zhou Z, Reiter RJ. Melatonin as a Potent and Inducible Endogenous Antioxidant: Synthesis and Metabolism. Molecules. 2015;20(10):18886–18906. doi:https://doi.org/10.3390/molecules201018886.

The Human Microbiome Project Consortium. Structure, function and diversity of the healthy human microbiome. Nature. 2012;486(7402):207–214. doi:https://doi.org/10.1038/nature11234.

Thellier M, Legent G, Amar P, Norris V, Ripoll C. Steady-state kinetic behavior of functioning-dependent structures. The FEBS journal. 2006;273(18):4287–4299. doi:https://doi.org/10.1111/j.1742-4658.2006.05425.x.

Thompson JA, Oliveira RA, Djukovic A, Ubeda C, Xavier KB. Manipulation of the Quorum-Sensing Signal AI-2 Affects the Antibiotic-Treated Gut Microbiota. Cell Reports. 2015;10(11):1861–71. doi:https://doi.org/10.1016/j.celrep.2015.02.049.

Tkachenko EI. The paradigm of dysbiosis in modern gastroenterology. The role of microbiota in the treatment and prevention of diseases in the XXI Century. Experimental & clinical gastroenterology. 2014;105(5):4–7 (in Russian).

Tomas-Barberan F, García-Villalba R, Quartieri A et al. In vitro transformation of chlorogenic acid by human gut microbiota. Mol Nutr Food Res. 2014;58(5):1122–1131. doi:https://doi.org/10.1002/mnfr.201300441.

Tomás-Barberán FA, González-Sarrías A, García-Villalba R et al. Urolithins, the rescue of «old» metabolites to understand a «new» concept: Metabotypes as a nexus among phenolic metabolism, microbiota dysbiosis, and host health status. Mol Nutr Food Res. 2017;61(1). doi:https://doi.org/10.1002/mnfr.201500901.

Trugnan G, Freitas M, Sapin C. Molecular basis for the cross talk between pathogens, intestinal cells, and gut microflora. In: Heidt PJ, Midtvedt T et al., editors. Old Herborn University Seminar Monograph. 15: Probiotics: bacteria and bacterial fragments as immunomodulatory agents. Germany: Herborn Litterae, Herborn-Dill; 2002. p. 15–23. https://www.old-herborn-university.de/wp-content/uploads/publications/books/OHUni_book_15_article_3.pdf. Accessed 28 June 2019.

Tsang AH, Barclay JL, Oster H. Interactions between endocrine and circadian systems. J Mol Endocrinol. 2013;52(1):R1–16. doi:https://doi.org/10.1530/JME-13-0118.

Tsilingiri K, Rescigno M. Postbiotics: what else? Benefic Microb. 2012;4(1):101–107. doi:https://doi.org/10.3920/BM2012.0046.

Tsuda T, Tsuda M, Yamamura V, Ohnishi N. Regulation of appetite and gastrointestinal microflora. J Germfree Life Gnotobiol. 1992;22:23–25.

Ugolev AM. Theory of adequate nutrition and trophology. Leningrad: Nauka; 1991. (in Russian).

Vakhitov TYa, Petrov LN, Bondarenko VM. The concept of a probiotic drug containing original microbial metabolites. Zh Mikrobiol (Moscow). 2005; 5:108–114 (in Russian).

Vakhitov TYa, Sitkin SI. The concept of superorganism in biology and medicine. Experimental & clinical gastroenterology. 2014;107(7):72–85 (in Russian).

Van Reenen CA, Dicks LM. Horizontal gene transfer amongst probiotic lactic acid bacteria and other intestinal microbiota: what are the possibilities? A review. Arch Microbiol. 2011;193(3):157–168. doi:https://doi.org/10.1007/s00203-010-0668-3.

Venema K, do Carmo AP. Probiotic and Prebiotics: Current Research and Future Trends. Wageningen: Caiser Academic Press; 2015. doi:https://doi.org/10.21775/9781910190098.

Verbeke F, De Craemer S, Debunne N, Janssens Y, Wynendaele E. Peptides as Quorum Sensing Molecules: Measurement Techniques and Obtained Levels In vitro and In vivo. Front Neurosci. 2017;11:183. doi:https://doi.org/10.3389/fnins.2017.00183.

Viganò L, Cinzia C, Oliveira A, Guerrieri P. *Helicobacter pylori*: Is there an Association with Oral Pathologies? A Traditional Review. Acta Scientific Microbiology. 2018;1(9):32–39.

Volkov MYu, Tkachenko EI, Vorobeichikov EV, Sinitsa AV. Bacillus subtilis metabolites as a novel promising probiotic preparations. Zh Mikrobiol (Moscow). 2007;2:75–80 (in Russian).

Volkov MYu, Zakharchenko MM, Sinitsa AV, Tkachenko EA, Uspenskiy YuP. Novel functional food products based on biotechnology. Military medical journal. 2006;5:69–70 (in Russian).

Volkov MYu. Toxicological studies of a new probiotic based on metabolites of bacilli and natural zeolite. News of higher educational institutions. North Caucasus region. Natural science. Special issue. 2008;23–26 (in Russian).

Voreades N, Kozil A, Weir TL. Diet and the development of the human intestinal microbiome. Front Microbiol. 2014;5:1–9. doi:https://doi.org/10.3389/fmicb.2014.00494.

Vorobeichikov EV, Stepanov AV, Volkov MYu, Vasilenko AZh, Ponomarenko VM, Sinitsa AV. Immunotropic effects of probiotic complex Bactistatin on the background of antibiotics. Antibiotics and chemotherapy. 2008;53(1):3–9 (in Russian).

Vyas U, Ranganathan N. Probiotics, prebiotics, and synbiotics: gut and beyond. Gastroenterology Research and Practice. 2012;2012:872716. doi:https://doi.org/10.1155/2012/872716.

Wallace BD, Redinbo MT. The Human Microbiome is a Source of Therapeutic Drug Targets. Curr Opin Chem Biol. 2013;17(3):379–384. doi:https://doi.org/10.1016/j.cbpa.2013.04.011.

Wallace DC. The epigenome and the mitochondrion: bioenergetics and the environment. Gen Dev. 2010;24(15):1571–1573. doi:https://doi.org/10.1101/gad.1960210.

Wang J, Li D, Dangott LJ, Wu G. Proteomics and its role in nutritional research. J Nutr. 2006;136(7):1759–1762. doi:https://doi.org/10.1093/jn/136.7.1759.

Wang Y, Wu Y, Wang Y, Xu H, Mei X et al. Antioxidant Properties of Probiotic Bacteria. Nutrients. 2017a;9(5):521. doi:https://doi.org/10.3390/nu9050521.

Wang Z, Klipfell E, Bennett BJ et al. Gut flora metabolism of phosphatidylcholine promotes cardiovascular disease. Nature. 2011;472(7341):57–63. doi:https://doi.org/10.1038/nature09922.

Wang ZJ, Chen XF, Zhang ZX, Li YC, Deng J, Tu J et al. Effects of anti-Helicobacter pylori concomitant therapy and probiotic supplementation on the throat and gut microbiota in humans. Microb Pathogenesis. 2017b;109:156–161. doi:https://doi.org/10.1016/j.micpath.2017.05.035.

Wikoff WR, Anfora AT, Liu J, Schultz PG, Lesley SA, Peters EC et al. Metabolomics analysis reveals large effects of gut microflora on mammalian blood metabolites. Proc Natl Acad Sci USA. 2009;106(10):3698–3703. doi:https://doi.org/10.1073/pnas.0812874106.

Wilson ID, Nicholson JK. Gut microbiome interaction with drug metabolism, efficacy and toxicity. Transl Res. 2017;179:204–222. doi:https://doi.org/10.1016/j.trsl.2016.08.002.

Yadav M, Verma MK, Chauhan NS. A review of metabolic potential of human gut microbiome in human nutrition. Arch Microbiol. 2018;200(2):203–217. doi:https://doi.org/10.1007/s00203-017-1459-x.

Yaglov VV, Yaglova NV. New concepts of diffuse endocrine system biology: results and prospects of its study. Herald of the Russian Academy of medical Sciences. 2012;4:74–81 (in Russian).

Yazdankhah SP, Midtvedt T, Narvhus JA, Berstad A, Lassen J, Halvorsen R. The use of probiotics for critically ill patients in hospitals. Microb Ecol Health Dis. 2009;21(3–4):114–121. doi:https://doi.org/10.3109/08910600903495046.

Yuan J, Wang B, Sun Z, Bo X, Yuan X et al. Analysis of host-inducing proteome changes in *Bifidobacterium longum* NCC2705 grown in vivo. J Proteome Res. 2008;7:375–385. doi:https://doi.org/10.1021/pr0704940.

Zarrilli R, Ricci V, Romano M. Molecular response of gastric epithelial cells to Helicobacter pylori-induced cell damage. Cell Microbiol. 1999;1(2):93–99.

Zelante T, Iannitti RG, Cunha C et al. Tryptophan catabolites from microbiota engage aryl hydrocarbon receptor and balance mucosal reactivity via interleukin-22. Immunity. 2013;39(2):372–385. doi:https://doi.org/10.1016/j.immuni.2013.08.003.

Zhernakova A, Kurilshikov A, Bonder MJ, Tigchelaar EF, Schirmer M et al. Population-based metagenomics analysis reveals markers for gut microbiome composition and diversity. Science. 2016;352(6285):565–569. doi:https://doi.org/10.1126/science.aad3369.

Zierer J, Jackson MA, Kastenmuller G, Maqngino M, Long T, Telenti A et al. The fecal metabolome as a functional readout of the gut microbiome. Nature genetics. 2018;50:790–795. doi:https://doi.org/10.1038/s41588-018-0135-7.

Zihler A, Le Blay G, de Wouters T, Lacroix C, Braegger CP et al. In vitro inhibition activity of defferent bacteriocin-producing *Escherichia coli* against *Salmonella* strains isolated from clinical cases. Letters in Applied Microbiology. 2009;49(1):31–38. doi:https://doi.org/10.1111/j.1472-765X.2009.02614.x.

Index

B. A. Shenderov et al., *METABIOTICS*,
https://doi.org/10.1007/978-3-030-34167-1

G

H

I

L

Printed by Printforce, the Netherlands